# INTRODUCTION TO

# FOOD CHEMISTRY

# INTRODUCTION TO
# FOOD CHEMISTRY

## richard owusu-apenten

**CRC Press**
Taylor & Francis Group
Boca Raton  London  New York

CRC Press is an imprint of the
Taylor & Francis Group, an **informa** business

Cover image courtesy of Penn State's College of Agricultural Sciences.

CRC Press
Taylor & Francis Group
6000 Broken Sound Parkway NW, Suite 300
Boca Raton, FL 33487-2742

First issued in paperback 2019

ISBN-13: 978-0-8493-1724-8 (hbk)
ISBN-13: 978-0-367-39348-9 (pbk)
Library of Congress Card Number 2004054513

### Library of Congress Cataloging-in-Publication Data

Owusu-Apenten, R. K.
   Introduction to food chemistry / Richard Owusu-Apenten.
     p. cm.
   Includes bibliographical references and index.
   ISBN 0-8493-1724-X
    1. Food-Analysis. I. Title.

TX545.O98 2004
664'.07--dc22

2004054513

**Visit the Taylor & Francis Web site at
http://www.taylorandfrancis.com**

**and the CRC Press Web site at
http://www.crcpress.com**

To Wenli

# Preface

Food chemistry is the application of chemistry principles to the food system, including agricultural production, storage, transportation, processing, distribution, retail, and consumption. The mission statement for food chemists is to ensure a supply of food that is safe, nutritious, and affordable, with due regard for the environment. Food chemistry is a challenging subject for at least three reasons. *First*, students of food chemistry need a good grounding in general chemistry. *Second*, they must possess the will to apply their chemistry knowledge to foods, which are invariably highly complex systems. *Third*, there is generally only a short time between discovery, integration, education, and application of new food chemistry ideas.

There are several advanced food chemistry books on the market, but there are very few *introductory* texts for food chemists. This book attempts to fill this niche. There is a balance of facts and explanations. Each chapter contains material for about 5 hours of reading followed by appropriate homework. Typically, food chemistry students have previous experience of college chemistry (organic and physical) and some calculus. They are computer literate and able to use the Internet for supplementary reading.

No previous training *in food chemistry* is assumed. Familiarity with and/or a willingness to engage some major themes in general chemistry (Chapter 1) is needed. **Introduction to Food Chemistry** is a multi-level text with material for a diverse audience. Within each of the twelve chapters, more complex ideas appear near the end. The book will be suitable as a basis for courses in food science with emphasis on food chemistry. Material from some parts of the book could be useful for nutrition majors. Another aim of this book is to help the reader proceed to more specialist monographs and research articles. Members of the food professions will also benefit from reading this book.

# Acknowledgments

*"It takes a village to raise one child."* I am indebted to many people who contributed directly or indirectly to this book. My sincere thanks to Professor D. S. Robinson, Professor M. Povey, and Professor B. Wedzicha for their inspiration. I thank Dr. Yannis Zebatakis, Athens University, for his friendship. Special regards to the many hundreds of students who took the courses FDSC 1020, FDSC 2030, FDSC 3040, and FDSC 5030 at the University of Leeds, UK. Gratitude is due to many present and past food science students at PSU who took FDSC 200, FDSC 201, FDSC 500, FDSC 508, and or FDSC 509. Without your participation this book would not have been written. I am grateful to my editor, Susan Farmer, and the staff of CRC Press for their help. Thanks are also given to my current boss, John Floros, for putting up with my "moonlighting." Finally, I thank my Ma and Grandma for their lasting love and patience.

Richard Owusu-Apenten
State College, PA
August 2004

# Author

**Richard Owusu-Apenten BSc, PhD, CChem, MRSC, MIFT** graduated with biochemistry (BSc with honors) and chemistry (PhD) degrees from the University of London. He has taught food chemistry for 15 years in universities in the UK and USA. The author is a chartered chemist and a professional member of the Royal Society of Chemistry, Institute of Food Technology, and the American Chemical Society. Dr. Apenten is an associate professor at the College of Agriculture, Pennsylvania State University.

# Table of Contents

# 1

# Chemistry and Food Chemistry: An Overview

## 1.1  INTRODUCTION

The study of chemistry probably dates from the first millennium AD, and the route to modern chemistry can be traced through the alchemist philosophies of medieval China, India, and Europe, although it took about a thousand years for chemistry to grow into a coherent and inclusive discipline.[1] The marriage of chemistry and biology led to many progeny, including agricultural chemistry, biological chemistry, clinical chemistry, forensic chemistry, industrial chemistry, pharmaceutical chemistry, and food chemistry. These branches of chemistry deal with economically significant issues such as agricultural production, health, crime, fuel processing, and the development of new medicines. Food chemistry deals with food and sustenance.

The aim of this chapter is to present an overview of chemistry (Sections 1.2 to 1.4). We also define food chemistry and explore its relation to the other *food sciences*, such as food analysis, microbiology, nutrition, and food engineering (Section 1.5). The excursion provides a check list of some of the major themes from food chemistry. Section 1.6 describes the organization of this book and deals with some anticipated learning outcomes and attainment goals.

The style for this book is partly a response to some emerging new trends in college education. For the foreseeable future, this seems to be the era of decreasing unit funding per student and smaller class sizes. There is increasing emphasis on improving learning and teaching (L&T) quality, measuring student based learning outcomes, and improving core competencies. Teaching loads for most faculty are increasing, and significant numbers of students are working to fund their studies. All this points to a need for better efficiency, perhaps through improved study skills, better teaching methods, and increased distance learning. After 15 years of teaching food chemistry it is noticeable that these trends bring with them constraints and challenges for the student and taxpayer. Additionally, we should not overlook some of the new opportunities offered by improving technologies and living standards.

### 1.1.1  ON THE NEED FOR CHEMISTRY

Chemistry is the study of the composition, structure, and properties of *materials* and the changes that these undergo. Inserting the word *food* before *materials* leads to a reasonable definition for food chemistry. In other words, food chemistry is the study of composition, structure, and properties of *food materials* and the changes that these undergo. A more general definition for food chemistry is presented later.

Food chemistry courses assume previous knowledge of chemistry, and students of food chemistry require a good foundation in general chemistry. The greater one's familiarity with chemistry, the greater the likelihood of success in food chemistry. Actually, the "chemistry issue" remains troubling to many food scientists including students, educators, and industrialists. There is no agreement on the *minimum* amount of chemistry needed for a successful food science degree. It is possible to study food chemistry *and* chemistry simultaneously so that the former is not an absolute *prerequisite*. Chemistry is normally divided into three parts: organic, physical, and inorganic. This chapter will provide a brief survey of the three main areas of chemistry. My aim is also to introduce some of the vocabulary and concepts used in later chapters of this book.

## 1.2 ORGANIC CHEMISTRY

The focus of organic chemistry is the compounds of carbon (Table 1.1). A common starting point is to consider **covalent** bonding between carbon and hydrogen. The naming convention for **hydrocarbons** containing up to ten carbons is then introduced as an example of organic **nomenclature:** this is the practice of giving the thousands of organic compounds different names. No two organic compounds have the same **systematic** name. For instance, the family of simple hydrocarbons called **alkanes** includes methane, ethane, propane, butane, pentane, and hexane, which have one, two, three, four, five or six carbon atoms, respectively. Structures of some simple hydrocarbons are shown in Figure 1.1.

After dealing with hydrocarbon structure, initial theories of bonding and **valence** theory are then introduced. A fully **saturated** carbon atom is bonded to four other atoms and is therefore described as tetravalent. Unsaturated hydrocarbons which have one or more carbon–carbon double or triple bonds are termed **alkenes** and **alkynes**, respectively. Considering **conformational isomers** formed by rotation at carbon–carbon single bonds and the **gauche** and **eclipsed** forms in ethane introduces the notion of the **steric** constraints for molecular conformation (Chapter 7)[a].

---

**Table 1.1**
**Organic chemistry: the study of carbon containing compounds**

1. Fundamentals of valence theory as applied to carbon
2. Functional group chemistry
   a. Hydrocarbons
   b. Alcohols
   c. Ethers, aldehydes, and ketones
   d. Carboxylic acid and esters
   e. Amines and amides
   f. Aromatic compounds
3. Polymers (synthetic and natural)
4. Natural product chemistry, biological chemistry
5. Organic reaction mechanisms
6. Stereochemistry
7. Organic synthesis

---

[a]See Figure 7.3 for further discussions on how steric factors determine the conformation of large molecules.

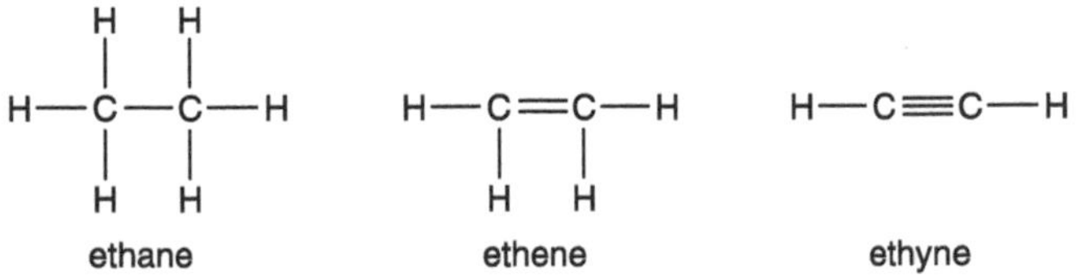

**Figure 1.1** The structure of some simple hydrocarbons. Ethane, ethene (ethylene), and ethyne (acetylene) are examples of alkanes, alkenes, and alkynes.

### 1.2.1 FUNCTIONAL GROUP CHEMISTRY

The many thousands of organic compounds belong to a limited number of families. Alkanes, alkenes, and alkynes are examples of chemical families described previously. Members of the **alcohol family** possess the general structure R–OH, where R is a hydrocarbon group linked to a hydroxyl group (–OH). The OH group (oxygen linked to a hydrogen atom) determines the types of reactions possible for alcohols. Other families of organic compounds include **ethers** (R–O–R), **aldehydes** (R–CHO), **ketones** (R.C=O.R), and **carboxylic acids** (R–COOH). Some organic functional groups and example compounds are shown in Figure 1.2.

The behavior of particular functional groups remain essentially unchanged wherever we find that group. For instance, the OH (hydroxyl) group behaves similarly when attached to methane, ethane, propane, etc. In other words, organic functional group chemistry is portable. We can apply our knowledge of alcohol group chemistry wherever we encounter this group. The same principles apply to the aldehyde, ketone, carboxylic acid and other groups. It is worth noting that particular functional group chemistries are moderated by the nature of the R group. For instance, ethyl alcohol is not very acidic whereas phenol (another alcohol) is a relatively strong acid. The structures of the some of the important functional groups and example compounds are shown in Figure 1.2. Discussions of functional group chemistry usually end with carbohydrates, proteins, and deoxyribonucleic acids as **natural products**.

### 1.2.2 AROMATIC COMPOUNDS

Benzene provides an example of **aromaticity**, electron **delocalization**, **hybridization**, and **resonance**. Some of these themes apply also to straight chain compounds possessing alternating single and double bonds. Unsaturated lipids and the fat-soluble vitamins contain **conjugated double** bonds, which accounts for their **color** and susceptibility to **oxidation** and photochemical damage (Chapter 9). Free radical chain reactions are usually discussed in connection with **polymerization** of natural monomers (isoprene units) to form rubber.

### 1.2.3 ORGANIC REACTION MECHANISMS

More advanced organic chemistry courses deal with **organic reaction mechanisms**, including *addition*, *elimination*, *substitution*, and *rearrangement* reactions. Many of these reactions occur in raw foods and more so during processing (see Chapter 10). In classical organic chemistry free radical reactions (involving the steps of *initiation*, *propagation*, and *termination*) are generally described with respect to the formation of rubber and synthetic polymers. Away from polymerization chemistry, free radical mechanism accounts for some spectacular processes in the life sciences, including lipid oxidation, photochemical processes, and many deteriorative changes, including aging.

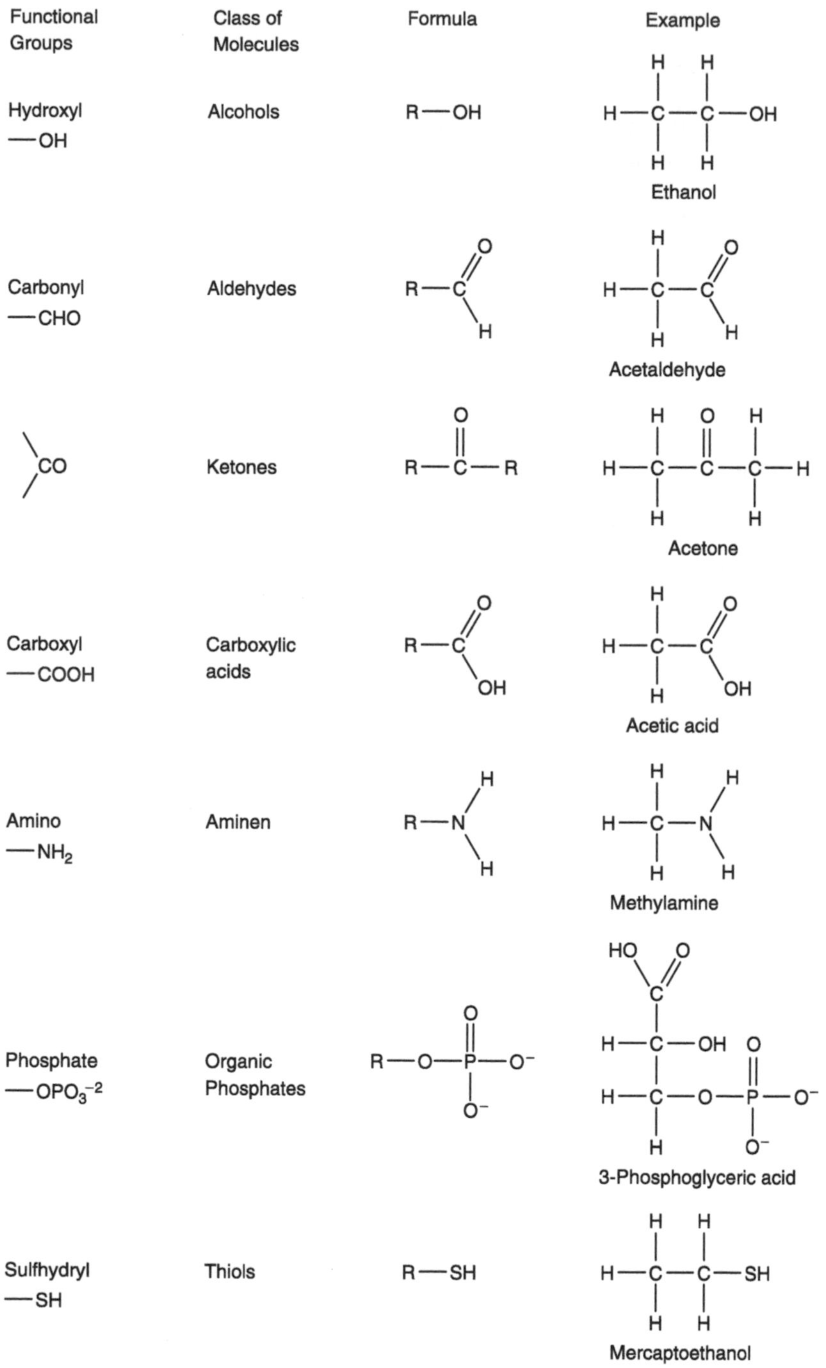

| Functional Groups | Class of Molecules | Formula | Example |
|---|---|---|---|

Hydroxyl —OH — Alcohols — R—OH — Ethanol

Carbonyl —CHO — Aldehydes — Acetaldehyde

>CO — Ketones — Acetone

Carboxyl —COOH — Carboxylic acids — Acetic acid

Amino —$NH_2$ — Aminen — Methylamine

Phosphate —$OPO_3^{-2}$ — Organic Phosphates — 3-Phosphoglyceric acid

Sulfhydryl —SH — Thiols — R—SH — Mercaptoethanol

**Figure 1.2** Some organic functional groups and example compounds.

## 1.2.4 STEREOCHEMISTRY

This branch of organic chemistry is concerned with compounds that possess *chiral* character. A good indication of chirality is that a compound contains an asymmetric carbon atom or **stereogenic center.** This is a carbon atom which is attached to four *different* groups.

Every additional asymmetric center yields two additional nonsuperimposible structures or **enantiomers** and in accordance with Van't Hoff's rule, there are $2^N$ stereoisomers for a molecule having $N$ asymmetric centers. Structural isomers that are not mirror images are called diastereoisomers or **diasteriomers**.

An asymmetric carbon is not entirely necessary for chirality. Large biological molecules possessing helices show chirality though they may lack a stereogenic center. The helix conformation cannot be superimposed on its mirror image. Chiral compounds exhibit **optical activity**, which is the ability to rotate plane-polarized light. A chiral molecule and its mirror image form a **recemic mixture** without net optical activity because each **enantiomer** has an opposite but equal effect on plane polarized light.

## 1.3   PHYSICAL CHEMISTRY

Physical chemistry describes material transformations. The three main themes of physical chemistry are thermodynamics, chemical kinetics, and quantum mechanics (Table 1.2).

### 1.3.1   THERMODYNAMICS

In eighteenth century Europe there was a great deal of interest in the nature of heat, work, and energy, and how to measure these attributes. Engineers formulated such questions during the Industrial Revolution, which led to the invention of the steam engine and early machinery. **Thermodynamics** (Greek: *therme*, heat; *dynamis*, *power*) is essentially the study of different forms of energy and the efficiencies with which those forms can be interconverted. There is emphasis also on how "energy flow" affects physico-chemical transformations.

The **first law of thermodynamics** states that energy is conserved—being neither destroyed nor created. A balance sheet can be drawn showing the conversion of mechanical energy to heat and other forms of energy. The total internal energy for a system ($U$) is available as heat ($Q$) or

---

**Table 1.2**
**Physical chemistry is the application of physical principles to chemical problems**

1. Thermodynamics
    a. First law (conservation of energy)
    b. Heat and work (thermochemistry, bond energies)
    c. Second and third laws of thermodynamics (concerning spontaneity, entropy, Gibbs free energy)
2. Chemical kinetics—discussing rates of change
    a. Rates of reaction, reaction order
    b. Determination of rate constants
    c. Effect of temperature on rates of reaction
3. Quantum chemistry
    a. Molecular orbital theory, chemical bonding
    b. Electron interactions with electromagnetic radiation
    c. Spectroscopy
4. Other areas
    a. Gas, liquid, and solid states
    b. Solutions (nonelectrolyte and electrolyte solutions)
    c. Chemical equilibrium
    d. Acid–base equilibria
    e. Electrochemistry
    f. Macromolecules

work ($W$). The idea is succinctly stated by the equation $U = Q + W$. The same relation is expressed in a slightly different way, $\Delta H = \Delta U + W$, for those transformations that take place under constant pressure conditions. $\Delta H$ is the **enthalpy change**, which is the heat absorbed or evolved by the reaction and $W$ is now the pressure–volume work or the amount of energy expended due to a volume change. By agreed convention, $\Delta H$ is given a negative sign for an exothermic reaction where heat is evolved. $\Delta H$ is positive for **endothermic reactions** where heat is absorbed. For most reactions taking place in solution, there is little or no volume change.

For a given reaction, the enthalpy change is equal to the heat contained in the reactants minus the heat contained in the products, in the form of chemical bonds. The magnitude of $\Delta H$ reflects the number of bonds broken minus the number of bonds re-formed and the relative strengths of **each**. Heats of reaction can be determined directly using **reaction calorimetry**. The quantity of chemical energy within different **food** components can be determined using **bomb calorimetry**. To summarize, the first law of thermodynamics and the associated discussions about the nature of "heat energy" provide us with the concepts of enthalpy, heats of combustion, as a measure of the energy content in different molecules, and bond energy. Figure 1.3 gives examples of heats of reaction.

The second law of thermodynamics states that changes occur so that the total **entropy** (the degree of disorder) available to the system increases. Ordinary material transformations occur in ways which lead to an increase in the amount of randomness in the system (see Appendix 1.1). Therefore entropy (as well as enthalpy) needs to be considered in order to ascertain whether a given transition will occur spontaneously. Combining the attributes of enthalpy and entropy yields the **Gibbs free energy** change ($\Delta G$) which is considered a wholly reliable indicator for **spontaneous** reactions. Figure 1.4 shows the Gibbs free energy change for transition between two states.

For many people, the Gibbs free energy change is *the* "big idea" in physical chemistry. This concept is applicable to a whole range of **reversible** processes between two final states, e.g., reactant and product, two concentrations of compounds, phase transformations from liquid to gas or vice versa, and also to conformations involving large molecules, like proteins. Consider the dilution process when water is added to a concentrated solution. Mixing one compound with another might not lead to a huge change in the number of bonds formed or broken. Nevertheless, dilution is a spontaneous process because the Gibbs free energy change for dilution is negative. The Gibbs free energy associated with an ensemble of molecules within a mixture is called the **chemical potential** or the partial molar Gibbs free energy.

---

Reactant → Product
$\Delta H = \sum \Delta H \text{ (products)} - \sum \Delta H \text{ (reactants)}$
$\Delta H = \sum (\text{ Number of bonds broken}) - \sum (\text{Number of bonds reformed})$

---

**Figure 1.3** Heats of reaction. The net heat change is due to the breaking and reformation of bonds including interactions with solvent.

---

Initial State ⇌ Final State
$\Delta G = \sum \Delta G \text{ (final state )} - \sum \Delta G \text{ (Initial state )}$
$\Delta G = \sum (\text{bonds broken } + \text{ disorder}) - \sum (\text{bonds reformed } + \text{ disorder})$

---

**Figure 1.4** Gibbs free energy change for transition between two states. A negative free energy change means the forward reaction will occur spontaneously.

Transformations leading to a net decrease in the chemical potential are spontaneous. (See Appendix 1.2.)

### 1.3.2 CHEMICAL KINETICS

Kinetics is the study of time-dependent phenomena. The rate with which things change is important in food chemistry and elsewhere. **Chemical kinetics** is probably the earliest form of kinetics encountered in general chemistry courses. However, principles of kinetics apply to all manner of physical, chemical, biological, and microbiological phenomena. It is perfectly normal to talk about the kinetics of (microbial) population growth, dissolution, and decay, for example. Chemical kinetics describes how rates of change depend on prevailing conditions, e.g., concentration of reacting compounds, temperature, pH, and ionic strength. The role of **catalysts** and inhibitors is also considered.

*The distinction between thermodynamics and kinetic control* is noteworthy. Thermodynamics tells us whether a process could occur, based on the known energy content of the starting and final states. Thermodynamics does not provide information related to rates of change. A thermodynamically favored process might take 1 second or a thousand years. In contrast, time-dependent changes of all kinds are described in a kinetics domain. The importance of kinetics cannot be overstated.

The rate of change is usually dependent on the concentration of reactant(s), [A] and [B], raised to a power ($x$, $y$, where $x$ and/or $y \geq 0$):

$$\text{rate} = k[\text{A}]^x[\text{B}]^y, \tag{1.1}$$

where $k$ is the **rate constant**. The **order of reaction** ($x + y$) is the number of molecules involved in the slowest (rate determining) step for a reaction. This so-called **molecularity** is also equal to the number of molecules needing to collide *simultaneously* in order for a reaction to occur. For $x = 0, 1, 2$ the reaction is said to be zero order, first order, or second order with respect to reactant A.

In **applied kinetics** we are interested in ways to determine the rate constant. Simple graphical methods are available for processing kinetics data in order to extract values for the rate constant and half-life ($t_{1/2}$) for a reaction[b]. The dependence of $k$ on temperature is described by the **Arrhenius equation**,

$$\ln k = \ln k_0 - \frac{\Delta E^{\#}}{RT}, \tag{1.2}$$

where $\Delta E^{\#}$ is the **activation energy** for reaction. $\Delta E^{\#}$ provides an indication of the **temperature sensitivity** of different reactions. Figure 1.5 shows (a) a typical Arrhenius graph in which the slope gives the activation energy; and (b) Svantes Arrhenius, after whom the equation is named.

The Arrhenius equation is extremely important for examining the dependence of $k$ on temperature. However, the scientific significance of the terms in this equation is controversial. According to the **collision theory**, reactions occur when two molecules collide, provided their combined energies exceed a minimum energy called the activation energy $\Delta E^{\#}$. Two colliding molecules with a combined energy exceeding $\Delta E^{\#}$ go on to form an **activated molecule**. This activated state then breaks down rapidly to form products. As an alternative to the collision

---

[b]Half-life and rate constant are directly related by the well known relation, $t_{1/2} = 0.69/k$.

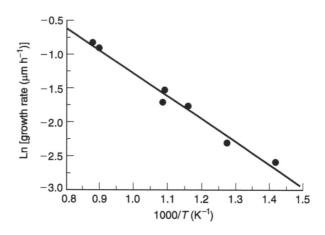

**Figure 1.5**  (Top) Arrhenius graph for the growth of microorganisms at different temperatures. The slope of this graph gives the activation energy for microbial growth. (From http://www.chm.bris.ac.uk/pt/diamond/image/arrhenius.gif.) (Bottom) Picture of Svantes Arrhenius, the Swedish scientist after whom the Arrhenius equation is named.

model, the **absolute reaction rate** theory of Eyring shows that $k$ is dependent on the activation *free* energy change ($\Delta G^{\#}$) needed to from a **transition state** (Eq. 1.3);

$$\Delta G^{\#} = -RT \ln\left(\frac{kh}{k_B T}\right), \tag{1.3}$$

$$\Delta G^{\#} = \Delta H^{\#} - T\Delta S^{\#}, \tag{1.4}$$

$$\Delta H^{\#} = \Delta E^{\#} - RT, \tag{1.5}$$

where $k_B$ is called the Boltzmann constant, $R$ is the universal gas constant and $h$ is Planck's constant. The formation of a transition state can be analyzed using thermodynamic terms such as the activation enthalpy ($\Delta H^{\#}$) change and activation entropy change ($\Delta S^{\#}$). In Eqs. 1.3 to 1.5, the hash sign (#) is obligatory to avoid confusion with equilibrium thermodynamic parameters described earlier (cf. Appendices 1.1 and 1.2).

We will end this brief survey of physical chemistry by drawing attention to the need for some understanding of theories of atomic structure, molecular bonding, and bond geometry. Quantum mechanics provides the theoretical framework for discussing spectroscopic analysis, which is a vital technique in food chemistry. Some of these ideas are described below.

**Table 1.3**
**Inorganic chemistry—reactions of inorganic compounds**

1. Periodic properties of elements other than carbon
    a. History of periodic table
    b Atomic radii, ionization energy, and electronegativity
    c. Metals, nonmetals, and metalloids
    d. Characteristics of alkali and alkali earth metals
    e. Group IIIA–VIIIA elements
2. Chemical bonding and molecular geometry
    a. Ionic, covalent, and metallic bonds
    b. Lattice energies
    c. Multiple bonds
    d. Electronegativity
    e. Lewis structures showing shared electrons
    f. Covalent bonding from orbital overlap
    g. Hybrid orbital
    h. Molecular orbital
3. Coordination chemistry
    Complexes and ligands
4. Bioinorganic chemistry (from cobalamine to heme proteins)

## 1.4   INORGANIC CHEMISTRY

Inorganic chemistry deals with all elements other than carbon, which includes metals, nonmetals, and metalloids. The 100 or so elements discovered so far can be described in terms of their positions in the periodic table developed by Mendeleev. Periodicity of **atomic radii**, and **ionization energies** of the elements depend on their **effective nuclear charge** and shielding. Variations in **ionization energies** lead to metal and nonmetal character. Regular periodicities in electron arrangements account for difference in **electronegativity** of different elements (Table 1.3).

### 1.4.1   CHEMICAL BONDING

Chemical bonding is readily described with the aid of **Lewis structures**. In the Lewis presentation two atoms bonding together are drawn as circles. Black dots are placed round each element showing the number of electrons in its outermost orbital. According to the **octet rule,** covalent bond forms such that each of two bonding atoms attain a complement of eight electrons in their outermost electron orbital. Differences in electronegativity produce uneven sharing of electrons and **polarized** bonds. In extreme cases there is a complete transfer of electrons from one atom to another, leading to **ionic bonding.** For those molecular structures where simple Lewis structures cannot be drawn, **resonance** may be invoked; alternative Lewis structures can be drawn having the same arrangement of atoms but with differing arrangement of electrons. The real structure is then assumed to be intermediate between two **resonance forms**. Inorganic compounds showing resonance **hybridization** include nitrate, sulfate, and phosphate ions. As an example, resonance structures for the nitrate ion are shown in Figure 1.6.

### 1.4.2   THE SHAPES OF MOLECULES

The shapes of simple molecules can be explained by the **valence-shell electron pair repulsion** (VSEPR) model. Accordingly, the geometric shapes adopted by different molecules are explained in terms of the repulsion between *bonds* linked to central atom. For molecules with

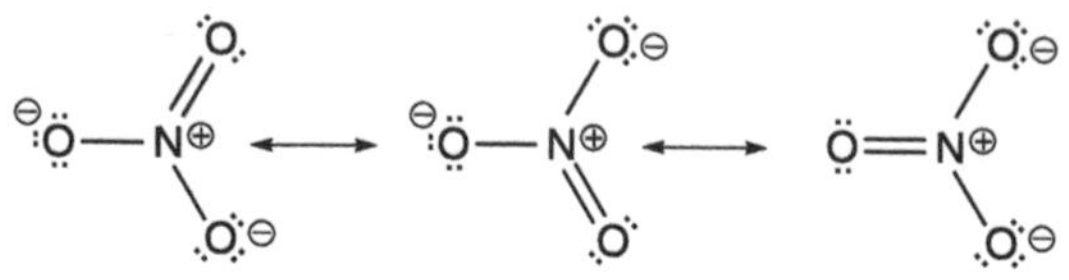

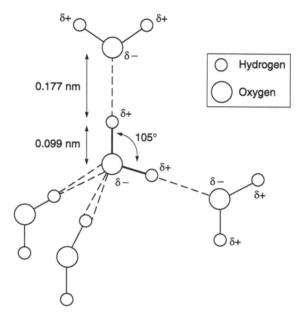

Figure 1.6  Resonance structures for the nitrate ion.

**Figure 1.7**  The structure of water.

the formulae $BX_2$, $BX_3$, $BX_4$, and $BX_5$ mutual repulsion between the X-groups will lead to linear, trigonal planar, tetrahedral, trigonal bipyrimidal or octahedral shapes that have the largest separations between the pendent groups. When the VSEPR model is applied to molecules such as **water** ($H_2O$) or ammonia ($NH_3$) any lone pair of electrons is treated as if they were bonds. The structure of water is shown in Figure 1.7.

### 1.4.3  VALENCE

The **valence-bond theory** provides a detailed description, as compared with Lewis structures, of covalent bond formation. Chemical bonds form due to the overlap of electronic orbitals. The valency of some elements is not always well explained by the number of single electrons in their outer shells. These instances can be explained by reference to **hybrid orbitals** which form in two steps: (1) energy input is needed to split and then promote one of a pair of electrons into a higher energy orbital, followed by (2) mixing of the resulting s, p or d orbitals to produce an *equivalent* set of $sp^1$, $sp^2$, $sp^3$ hybrid orbitals. The shapes of these orbitals are determined by mutual electron repulsion as described above.

### 1.4.4  MOLECULAR ORBITAL THEORY

The ideas of chemical bonding developed over several hundred years, culminating with the **molecular orbital** theory (see Appendix 1.3), key elements of which are that (1) atomic obitals can be combined to form molecular orbitals; (2) the number of molecular obitals formed is equal to the number of atomic orbitals combined; (3) upon combining atomic orbitals,

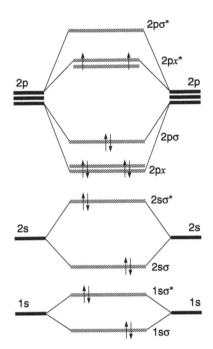

**Figure 1.8** The arrangement of molecular orbitals in diatomic oxygen molecule ($O_2$). Compare this diagram with the similar diagram in Chapter 9.

an equal number of **bonding** and **antibonding** orbitals are formed; (4) each molecular orbital can only accommodate two electrons of opposing spin (Pauli's exclusion principle); and (5) molecular orbitals are filled with single electrons before pairing occurs.

The molecular orbitals for diatomic oxygen are shown in Figure 1.8. Two sets of atomic orbitals for elemental oxygen ($1s^2, 2s^2, 2p^4$) are combined to produce molecular orbitals. Accordingly, the 1s orbitals from each of two oxygen atoms combine to produce two molecular orbitals, one of which is bonding ($1\sigma$) whereas the other is antibonding ($1\sigma^*$). This general rule is worth remembering: the overlap of $x$ pairs of atomic orbitals produces $2x$ molecular orbitals, one half of which are bonding and antibonding orbitals. The former have generally lower energy than the former. The total of 12 electrons in diatomic oxygen is arrayed in a manner that leads to the lowest energy arrangement.

The preceding sections represent a necessarily brief review of general chemistry. Readers should continue to cultivate their understanding of chemistry during their study of food chemistry. Two websites, one from MIT and the other from Purdue University are especially useful for reviewing general chemistry topics. A further useful resource is a site devoted to chemistry for nutrition science[c].

## 1.5  FOOD CHEMISTRY

### 1.5.1  Definition and Scope

Food chemistry is the application of **chemistry** principles to the **food system**. There is emphasis on the chemistry of food components, including macroconstituents (water, carbohydrates,

---

[c]http://chemed.chem.purdue.edu/genchem/topicreview/, http://web.mit.edu/esgbio/www/chem/review.html, and http://nutrition.jbpub.com/discovering/chemistry_review.cfm

lipids, and proteins), microconstituents (for example, flavors, vitamins, minerals, sweeteners, general additives), and their interactions. Food chemistry emerged as a discipline after World War II. The mission statement for food chemistry is to ensure a supply of food, which is nutritious, safe, and affordable, with due regard for the environment. This mission is shared by the other food sciences including food microbiology, food processing, food engineering, and food law.

After World War II, the emerging discipline of **food science** brought together a host of basic sciences (chemistry, mathematics, microbiology, and engineering) in a bid to improve the food supply. Food science (singular), the scientific study of foods and the food system, described in Section 1.5, is an assemblage of many sub-disciplines or *food sciences*, including food chemistry, food engineering, food microbiology, and food law. At first there was a clear emphasis on food processing and the **chemical transformations** encountered during processing. Advances were made in understanding the Maillard reaction, which leads to colored and flavored compounds during processing. The effect of moisture content on food deterioration was also examined leading to improvements in shelf-life of stored foods. Advances in synthetic organic chemistry led to a new range of **plastics** and **packaging** materials that transformed the food retail sector as recently as the 1980s. The behavior of food hydrocolloids, including starch, pectin, marine polysaccharides, and the protein gelatin, were elucidated. Developments in **food analysis** enabled the detection of pesticide residues and other toxicants in foods.

### 1.5.2  AREAS OF EXPERTISE REQUIRED BY THE INSTITUTE OF FOOD TECHNOLOGY

There are at least five core competencies for food scientists: **chemistry**, analysis, nutrition, microbiology, and engineering (Table 1.4)[d]. Not surprisingly, food science degree programs in North America and Europe offer courses in food chemistry and some proficiency in this area is essential for courses approved by the Institute of Food Technology.

### 1.5.3  CHEMISTRY AND THE FOOD SYSTEM

Food chemistry is the chemistry of food components. Nowadays, this view is unnecessarily restrictive. Measuring soil pH is as much a part of food chemistry as examining the effect of pasteurization on milk enzymes. Chemistry is found at all levels in the food system.

**Table 1.4**
**Organizing themes in food science**

| IFT expertise areas | Others | The food system |
| --- | --- | --- |
| • Chemistry | • Law | • Agricultural production |
| • Analysis | • Economics | • Storage |
| • Nutrition | • Ethics | • Processing and packaging |
| • Microbiology | • Psychology | • Distribution and retail |
| • Engineering | | • Consumer |
| • Success skills | | |

[d]Adapted from the US Institute of Food Technology (IFT) website: http://www.ift.org.education standards.html

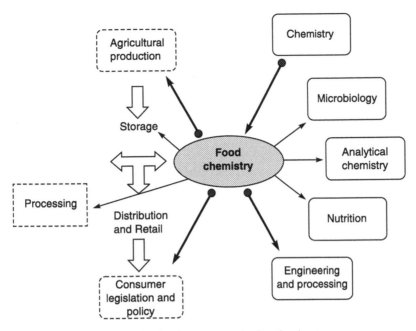

**Figure 1.9**   Relating food chemistry, the food system, and other food sciences.

Therefore, the scope of food chemistry is broad. We can show this graphically, by drawing lines from "chemistry" (Table 1.4) to each component of the food system. This exercise shows that food **chemistry** extends to agronomy, harvesting, extraction, processing and/or refining, packaging, storage, distribution, and retail (Figure 1.9). We need to "use" food chemistry also to study the behavior of food ingredients during food manufacture.

## 1.6   ORGANIZATION OF THIS BOOK

### 1.6.1   STRUCTURE AND RATIONALE

The structure of this book is different from those encountered in more advanced food chemistry texts,[2–4] as it begins with a review of chemistry and food chemistry. We feel this approach is justified for our intended audience, who could be expert in foods or chemistry but new to food chemistry. This first chapter is intended to get everyone "on the same page."

Chapters 2 and 3 are devoted to food analysis and statistical analysis. Food chemistry is an applied science. Careful measurement or analysis is essential. An understanding of statistics is necessary in order to decide whether those measurements are reliable. It is only through the proper use of statistical methods that we obtain reliable data and make substantive progress. Chapters 4 to 6 cover most of the conventional topics in food chemistry (e.g., carbohydrates, lipids, proteins, enzymes). Chapters 7 and 8 present **material science** and rheology, concepts that are being applied increasingly to foods. Together, these chapters provide a useful introduction to the measurement of texture and physical characteristics of foods. Chapters 9 and 10 consider chemical processes leading to food spoilage. Nonenzymic oxidation and the Maillard reaction are both important and interlinked. Finally, Chapters 11 and 12 deal with enzymic and biological chemistry of deteriorative processes, such as ripening and **senescence**.

### 1.6.2 AUDIENCE AND READERSHIP

Typical readers will have previous experience with college-level chemistry courses including organic, physical, and biological chemistry. They are computer literate and able to access the Internet for supplementary reading. The World Wide Web is an indispensable library resource if used carefully[e].

### 1.6.3 SPECIFIC OUTCOMES

This book aims to provide material for courses in food chemistry. In this section, we address the issue of student based outcomes. What specific skills and knowledge are you to expect from reading this text? Before making or seeking specific promises, consider two common readership scenarios. In scenario #1, you are reading *Introduction to Food Chemistry* as recommended by a course instructor. Clearly, studying the assigned sections *may* help you obtain a coveted A grade for your next class test. In readership scenario #2, you are interested in food chemistry for career or general reasons. Hopefully, this book will provide a short but stimulating introduction to food chemistry, allowing you to do your job more effectively.

Learning outcomes are currently a major concern. For instance, Bloom's taxonomy identifies three types of learning or *learning domains*. These are: (1) **cognitive domain**, related to knowledge and intellectual development; (2) **affective domain**, related to emotional development, feelings, values, enthusiasms, motivations, and attitudes; and (3) **psychomotor domain**, related to physical movement, coordination, and motor skills. It is easy to see that these three domains reflect some of the major themes for liberal arts education. Also interestingly, the cognitive domain is divided into *six levels of attainment:* (1) knowledge, (2) understanding, (3) application, (4) analysis, (5) synthesis, and (6) evaluation (Figure 1.10).[5] Depending on your viewpoint, knowledge is *merely* the lowest form of learning. Contrariwise, "We should not build castles on air." Consider knowledge as the fundamental element from which other forms of learning may emerge. Each cognitive domain or attainment level carries a complement of *demonstrable* skills and the means for their assessment.

After completing this book the reader will have learned a great many new facts intended to increase their knowledge of food chemistry. By emphasizing general principles, this book will facilitate an understanding of concepts (Figure 1.10). Though not adopting a commodities approach, we provide pertinent examples of areas of application for the concepts discussed. We hope that our intended audience will work towards comprehension and applications of chemistry principles to the food system. *Introduction to Food Chemistry* makes no specific claims regarding attainment targets located on the top three rungs of the cognitive domain pyramid.

## APPENDIX 1.1   STATISTICAL DESCRIPTION OF ENTROPY

In the statistical description of entropy developed by Ludwig Boltzmann, entropy ($S$) is related to the number of arrangements of atoms or molecules ($\omega$) possible in one state. $S = k_B \ln \omega$, where $k_B$ is the Boltzmann constant with a value $1.38 \times 10^{-23}$J/K. The Boltzmann entropy equation is engraved on his tomb in Vienna. Entropy can also be described using thermodynamic quantities, $\Delta S = \Delta H / T$.

---

[e]A time-line showing the history of the Internet can be found at www.w3.org/History.html. Commercial access to the WWW is a little more than 10 years old. Some material from the WWW has not undergone peer review and should be given the usual *healthy* skepticism.

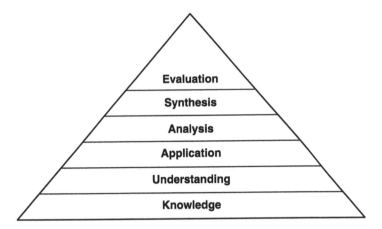

1. **Knowledge**: define, duplicate, label, list, memorize, name, order, recognize, relate, recall, repeat, reproduce, state.

2. **Understanding**: classify, describe, discuss, explain, identify, indicate, locate, recognize, report, restate, review, select, translate.

3. **Application**: apply, choose, demonstrate, dramatize, employ, illustrate, interpret, operate, practice, schedule, sketch, solve, use, write.

4. **Analysis**: analyze, appraise, calculate, categorize, compare, contrast, criticize, differentiate, discriminate, distinguish, examine, experiment, question, test.

5. **Synthesis**: arrange, assemble, collect, compose, construct, create, design, develop, formulate, manage, organize, plan, prepare, propose, set up, write.

6. **Evaluation**: appraise, argue, assess, attack, choose, compare, defend, estimate, judge, predict, rate, score, select, support, value, evaluate.

**Figure 1.10** Bloom's taxonomy: The six levels of attainment in the cognitive learning domain arranged as a pyramid. Boxed legend shows possible measurement criteria for each attainment target.

## APPENDIX 1.2  QUANTITY OF HEAT EVOLVED

The quantity of heat evolved provides some idea of the likelihood that a reaction will occur spontaneously, but enthalpy change is only one of the predictors of spontaneity. Put formally, discussions based on heat of formation assume that the entropy change for a reaction can be ignored. Approximately 150 years passed between the first law and the second and third laws of thermodynamics which introduced the idea of entropy and Gibbs free energy (named after Josiah Willard Gibbs) as predictors for spontaneous transformations. The following elementary relations are worth noting:

$$\Delta G = \Delta H - T\Delta S.$$

$\Delta G$ is negative for spontaneous reactions and positive for non-spontaneous reactions.

$$\Delta G = -RT\ln(K_{eq}) = -RT\ln\left(\frac{C_2}{C_1}\right),$$

where $K_{eq}$ is called the equilibrium constant. For transformation to occur spontaneously, $\Delta G$ must be negative. This condition is more likely when $\Delta S$ is positive (randomness increases) and/or when $\Delta H$ is negative (heat is evolved).

## APPENDIX 1.3  ORIGINS OF CHEMICAL BONDING

Chemical bonding theory has its origins in Dalton's theory that all matter is made of atoms that are indivisible. This was followed by Thompson's (plum pudding) model for the atom as a positively charged pudding with electrons embedded within it. Lord Rutherford, after irradiating gold leaves with alpha, beta, and gamma radiation proposed the "solar system" model for atoms with a central nucleus and orbiting electrons. Niels Bohr, from studies of emission spectra, suggested that electrons occupied orbitals with different energy levels and that inter-orbital transitions led to the emission of discrete (quantized) energy. Einstein studied the photoelectric effect and also proved that light was quantized. De Broglie suggested that electrons had energy and also wavelike properties. Heisenberg's uncertainty principle indicated that wave-matter, as characterized by electrons, could not be precisely located in space (or within orbitals) if the energy was known precisely. Schrödinger suggested that an atomic orbital was actually a region of space with a high probability of finding an electron. The currently accepted shapes for s, p, d, and f orbitals are 3-D representations of mathematical solutions to Schrödinger equations. Bonding results from the overlap of electron orbitals.

## References

1. Brown, T. L., Lemay, H. E., Burstein, B. E., and Burdge, J. R., *Chemistry: The Central Science* (9th edition), Prentice Hall, New York, 1, 2003.
2. De Man, J. M., *Principles of Food Chemistry*, Aspen Publishers Inc., Gaithesburg, Maryland, 1999.
3. Belitz, H.-D., Grosch, W. and Burghagen, M. M., *Food Chemistry*, Springer Verlag, Berlin, Germany, 1999.
4. Fennema, O. R., Ed., *Food Chemistry* (3rd edition), Marcel Dekker Inc., New York, 1996.
5. Adapted from http://www.eecs.usma.edu/cs383/bloom/default.htm

# 2

# Food Analysis

## 2.1 INTRODUCTION

Analysis may be defined as *the separation of anything into its parts; examination of something to determine its essential features; the intentional separation of substances into their ingredients or elements to find their amounts and nature; the determination of the kind or amount of one or more of the constituents of a substance whether actually obtained in separate form or not*. Food chemists employ a host of instrumental and other methods for analyzing food components. More generally, a host of analytical strategies may be applied to most components of the food system. Analytical procedures also involve **separation** and **measurement** stages.[1] Sensors and detectors are used in conjunction with information processing systems that convert analogue information (signals, observations) into digital (numerical) data.

This chapter provides an introduction to food analysis. We consider proximate analysis (Section 2.2) which refers to a cohort of techniques used to determine the basic constituents of foods (protein, fats, carbohydrates, fiber, total minerals as ash). Column chromatography methods (Section 2.3) are described, these being some of the most frequently used methods in food chemistry. Analysis based on the absorption of light is described in Section 2.4. The photometric techniques include simple color measurements using a colorimeter. Photometric methods also extend to the measurement of visible and ultraviolet spectra and infrared analysis.

## 2.1.1 ANALYTICAL CHEMISTRY OF FOODS

The result of quantitative analysis is **enumeration**. This process generates numerical data which is then interpreted using statistical principles (Chapter 3). Some common instrumental methods for quantitative analysis include photometry, electrophoresis, electrochemistry, and chromatography. Other techniques highlight sensorial qualities such as food texture (thickness, consistency, hardness, springiness). Texture analysis is described in the chapter on food rheology (Chapter 8). Color and flavor are the other important sensory attributes.

The food analyst is concerned with agricultural production, processing, transport and distribution, retail, sale, and consumption. At the consumer end of the spectrum analytical issues are related to **nutritional chemistry** and the determination of shelf-life and acceptable dates. Many techniques from clinical analysis (e.g., determination of vitamins, serum cholesterol, fecal protein or nitrogen) are also useful for food analysis. The major forms of food analysis are summarized in Table 2.1.

Living things possess sophisticated sensory apparatus for hearing, olfaction, taste, touch, and pressure sensing. These are analytical capabilities of enormous sensitivity and discriminatory power. There is considerable interest in instrumental methods for sensory analysis as surrogates for the sensory apparatus possessed by human beings, who are the ultimate

---

**Table 2.1**
**Forms of food analysis**

**Chemical methods** (wet chemistry methods)
Proximate analysis (water, fat, protein, carbohydrates, sugars, vitamins)
**Biological and biochemical methods**
Enzymatic methods
Microbiological methods
Whole animal assays (including human subjects)*
**Instrumental methods**
Spectrophotometry (absorption and emission methods)
Electroanalytical methods (potentiometry and voltametry)
Mass determination (gravimetric analysis, mass spectrometry)
Chromatographic methods
Thermal analysis (differential scanning calorimetry)
**Physical characterization**
Color measurement
Mechanical testing (texture, viscosity, rheology)
Particle sizing (sieving, gel permeation, light scattering)

*Includes feeding trials and sensory tests.

---

consumers of food products. Clearly, one hopes for correlation between analytical results obtained from classical, i.e., human based sensory analysis, and results obtained using mechanical instrumentation. Techniques of food analysis are continually undergoing **collaborative testing**. Those techniques which deliver an acceptable level of reliability receive "approved" status from organizations such as the Association of Official Analytical Chemists (AOAC). Using AOAC approved methods of analysis allows results to be compared meaningfully.

The measurement of fundamental physical parameters such as mass, volume, density, temperature, heat capacity, conductance, resistance, potential, electrical conductance, and impedance are all of interest to food scientists. With few exceptions, all numerical data found in tables, graphs, illustrations, and the conclusions drawn from such data, arose from some form of analysis.

## 2.2   PROXIMATE ANALYSIS

Proximate analysis refers to the analysis of foodstuffs to establish their composition. Proximate analysis provides information related to the content of water, protein, carbohydrates, lipids, fiber, and total minerals. Such information is necessary for labeling purposes and for quality control.

### 2.2.1   DETERMINATION OF WATER

The water content of many fresh foods ranges from 60 to 95%. The water content of dehydrated foodstuffs is kept vanishingly small in order to prolong shelf-life. The most common methods for water analysis are listed below.[2]

- Oven drying (98–100°C)
- Vacuum oven drying (60–70°C)
- Room temperature drying/vacuum dedicator
- Distillation with immiscible solvent

- Karl Fisher method
- Infrared analysis
- Gas chromatography

### 2.2.1.1  Oven Drying

Place 2–10 g (±1 mg) of sample in a flat-bottomed metallic flask pre-dried at 98°C for 60 min. Dry the sample by heating for times ranging from 2 to 3 h to overnight. Weigh the sample periodically until it reaches a constant weight (±2 mg). The percent moisture content can be calculated from the difference between the initial sample weight ($W_I$) and the final sample weight after drying ($W_F$):

$$\% \ moisture = \left(\frac{W_I - W_F}{W_1}\right) \times 100. \tag{2.1}$$

During oven drying, high sugar products should not be dried at greater than 70°C to avoid decomposition. At high temperatures, volatile components (flavors) can be lost by evaporation along with moisture. A more accurate alternative to oven drying is to use sample distillation. Lastly, care is needed to avoid rehydration of dried material from moist air. Samples should be covered and weighed as soon as possible after drying and cooling.

### 2.2.1.2  The Karl Fisher Method for Water Determination

In the presence of water, iodine reacts with other components of the Karl Fisher (KF) reagent to produce bound iodine. Once water has been consumed, excess iodine appears as a brown coloration. The titration endpoint can be determined visually or by coulometric analysis. The composition of the KF reagent is listed below. The first three components are mixed at room temperature and then $SO_2$ is added slowly to avoid overheating. The composition of the KF reagent is given in Table 2.2.

To perform KF moisture determination, a known quantity of the food material is dispersed in dry methanol and placed in a purpose made container fitted with an electrode (Figure 2.1). The instrument is then set up according to the manufacturer's instructions. Small increments of KF reagent are added with stirring. In the presence of water, components of the KF reagent react to produce the bound form of iodine:

$$H_2O + SO_2 + 3Pyr + I_2 \rightarrow Pyr \ sulfate + 2Pyr \ iodide. \tag{2.2}$$

As increasing amounts of the KF reagent are added, water in the methanolic dispersion is used up (Eq. 2.2) and free iodine suddenly appears. Free iodine can detected by electrolysis at two electrodes placed within the sample (Figure 2.1). A sudden flow of current, which occurs as free iodine reacts at the electrodes, indicates the titration endpoint. For uncolored samples, a visual endpoint can be seen as a brown coloration due to excess iodine.

It is necessary to calibrate the KF reagent in order to determine the calibration constant ($C_{KF}$): the volume of KF reagent required to react with a known amount of water added to

**Table 2.2**
**Composition of the Karl Fisher reagent**

| | |
|---|---|
| • Iodine | 133 g |
| • Pyridine (Pyr) | 425 g |
| • Methanol | 425 g |
| • Sulfur dioxide | 100 g |

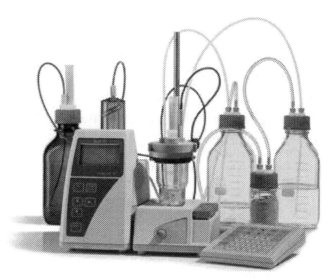

**Figure 2.1** A commercial Karl Fisher titrator. (From http://www.metrohm.com/products/02/795/795.html.)

dry methanol. An example application for the KF method is the determination of the moisture content of molasses or dry vegetables. From the volume of KF reagent ($V$) required to reach the endpoint, the moisture content of the sample is calculated from Eq. 2.3:

$$\% \ moisture = \left(\frac{C_{KF} \times V}{W}\right) \times 100, \tag{2.3}$$

where $W$ (mg) is the weight of sample analyzed. Typically, 1 ml of KF reagent reacts with about 3.5 mg of water, therefore the value of $C_{KF}$ is 3.5 mg/ml.

### 2.2.1.3 Other Methods for Water Determination

As described above, oven drying is not completely suited for samples that contain volatile components, such as fats, oils, waxes, cereals, and a wide range of plant materials and foodstuffs. Distillation is an alternative method for water determination which avoids errors arising from the loss of volatiles. The food sample is suspended in an organic solvent (e.g., toluene or xylene) that has a boiling point higher than the boiling point of water. Next, the suspension is heated so that water distils off as steam. This is collected using a water-cooled condenser, and measured. Other methods for water determination include gas chromatography and infrared analysis.

### 2.2.2 Determination of Crude Protein

#### 2.2.2.1 Kjeldahl Analysis

Kjeldahl analysis (named after the Danish chemist Johan Kjeldahl, Figure 2.2) is probably the best-known method for protein analysis. First, the food to be analyzed is treated with concentrated sulfuric acid at high temperatures. The acid converts all nitrogen compounds into ammonium sulfate. To yield accurate results, it is necessary for the sample to wholly decompose and oxidize to carbon dioxide, water, ammonia, and sulfur oxides. Metal oxide catalysts are usually added to the sulfuric acid to increase its boiling point to about 320°C, and to catalyze sample digestion.

**Figure 2.2** A portrait of Johan Gustav Christoffer Thorsagger Kjeldahl (1849–1900), Danish chemist and director of the Carlsberg laboratory, Copenhagen, Denmark, from 1876 to 1900, and inventor of the Kjeldahl method for protein nitrogen determination in biological samples. (Adapted from http://search. biography.com.)

Ammonium sulfate produced by acid digestion is neutralized with excess alkali and the resulting ammonia vapor is distilled and collected into a container of standard acid where it reacts. Thereafter, the quantity of standard acid remaining after the distillation phase can be determined by titration. The volume of titrant can be converted into the percentage of nitrogen (%N) contained in the initial sample. To calculate protein content the %N value is multiplied by a Kjeldahl factor. The default value for this constant is 6.25. The Kjeldahl method can be applied to solids, liquids, and slurries. The initial color of the sample is not important. The main sources of error are the presence of nonprotein nitrogen compounds associated with some foods. High concentrations of peptides, amino acids, nucleic acids or urea will give false positive results. Indeed, Kjeldahl analysis can be used to determine *nitrogen* levels in a wide variety of agricultural materials, ranging from fertilizer and feeds to coal. There is sometimes disagreement about the appropriate value for the Kjeldahl factor for different types of samples. Nitrogen from refractory material (nitrate compounds) is harder to determine.

### 2.2.2.2 Dumas or Combustion Analysis

Combustion nitrogen analyzers work on the principle developed by Dumas. A sample is burnt in a 99%+ oxygen atmosphere at 950–1000°C to form carbon dioxide, water, nitrogen oxide, and sulfur dioxide. The gases produced are absorbed using specific adsorbents, leaving only nitrogen oxide. This is first transformed to nitrogen and is then measured using a thermal conductivity detector. As with the Kjeldahl method, sample nitrogen is converted to protein by multiplying with a nitrogen-to-protein conversion factor. Combustion analysis has advantages over the Kjeldahl analysis. The former technique is quicker, safer, and uses more environmental friendly reagents.

### 2.2.3 CRUDE FAT

The crude fat content can be determined as the weight change recorded after exhaustively extracting a food substance with a nonpolar solvent. Materials removed by organic solvent extraction include triacylglycerols, phospholipids, sterols, and waxes. The instrument used for defatting samples is the **Soxhlet** extractor (Figure 2.3).

The Soxhlet extractor is a glass apparatus about 1 foot high and 1.5 inches in diameter (Figure 2.3). The extractor has two chambers. The top chamber contains the food material

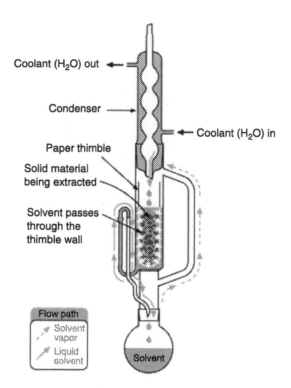

Coolant (H$_2$O) out

Condenser

Coolant (H$_2$O) in

Paper thimble

Solid material
being extracted

Solvent passes
through the
thimble wall

Flow path

Solvent
vapor

Liquid
solvent

Solvent

**Figure 2.3**   Diagram of a Soxhlet glass apparatus for defatting foodstuff using solvent extraction. A source of heat is not shown. (From http://www.anl.gov/OPA/logos16-2/extractor3.htm.)

with a fritted chamber that is fitted with a water jacketed condenser. The lower chamber contains a reservoir for organic solvent. Upon heating, the organic solvent vapor rises through a sidearm and condenses in the upper chamber whereupon it percolates downwards through the food. The extracted lipids fall back to the bottom reservoir where the solvent is revaporized. In this way the the food is continuously re-extracted by the circulating organic solvent.

### 2.2.4   ASH

The determination of ash provides a quick measure of the total content of minerals. Burning converts nonvolatiles into oxides and carbonates, which are then weighed.

### 2.2.5   CRUDE FIBER

Crude fiber usually consists of plant cell wall polysaccharides which are nondigestible by monogastric animals. Cellulose is a major constituent, along with hemicellulose, a polymer of xylanose, pectin, and some proteins (extensins). These so-called nonstarch polysaccharides function as fiber because humans lack the enzymes needed to degrade such polymers. Fiber is measured as the quantity of filterable solid remaining after a food sample is treated with 1.25% sulfuric acid followed by 1.25% sodium hydroxide. The sample is filtered after acid treatment and then filtered again after treatment with alkali. These processes are meant to simulate exposure of food samples to acidic and basic environments in the stomach and small intestines, respectively.

### 2.2.6 Digestible Carbohydrate

The carbohydrate content can be estimated as the weight of food material remaining after subtracting the weights for fiber, protein, lipid, and moisture.

## 2.3 COLUMN CHROMATOGRAPHIC ANALYSIS

### 2.3.1 Introduction

Chromatography is probably the single most useful form of chemical analysis used by food scientists.[3] The developments of chromatography over the past 100 years are summarized in Table 2.3. The most important chromatographic techniques employ columns for sample separation. Paper chromatography is another useful technique. Another arrangement involves finely divided support material adsorbed onto a glass plate for **thin-layer chromatography** (TLC).

A typical column chromatography system is shown in Figure 2.4. A flow-line carries solvent from a reservoir (1) via a pump (2) to an injection valve (3i); 3ii shows the point of sample injection into the flow-line carrying the **mobile phase**. This flows to a column (4) packed with an inert and finely divided support called a **stationary phase**. This consists of small diameter beads made from materials such as diatomaceous earth, cellulose, starch or glass.

Separation of sample components occurs during column chromatography due to repeated partitioning between the mobile and stationary phases. The number of notional partitioning steps within a column is called the **height equivalent theoretical plate**. To separate two components it is necessary for them to have different partitioning characteristics. For instance, a given column may separate two components due to differences in their electric charge, polarity or affinity for a ligand covalently attached to the column support. The separation process on the column leads to different sample components eluting/emerging from the column at different times.

These different components are carried to the **detector** (5) where their concentration is measured. A simple detector consists of a colorimeter fitted with a flow-through cell. The detector response, in the form of an electrical signal, passes to a strip-chart recorder or computer (7) for display and further data analysis. The recorded chromatographic profile is a series of peaks and troughs. Each peak normally corresponds to a single pure compound present in the original mixture.

**Column chromatography** systems differ according to the choice of (1) column support, (2) mobile phase, and (3) detector. With HPLC the column packing consists of small diameter glass beads. Small diameter packing material increases the sample resolution. Unfortunately, small diameter supports also produce a large pressure drop across a

**Table 2.3**
**Chronological developments in chromatography**

| Date | Comments |
|------|----------|
| 1903–1906 | Tswett—Separation of plant pigments by column chromatography |
| 1930–1932 | Karrer, Khum, and Strain—Use of activated lime as support |
| 1939 | Brown—Development of paper chromatography |
| 1940–1943 | Tselius—Development of thin-layer chromatography (TLC) |
| 1940 | Martin and Synge—Theory of chromatography |
| 1956 | Sober and Peterson—Cellulose support for ion exchange chromatography |
| 1970 | Development of first high-pressure liquid chromatography (HPLC) |

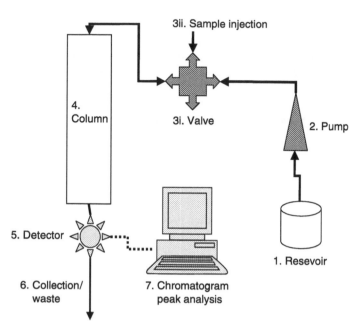

**Figure 2.4** Components of column chromatography.

**Table 2.4**
**The characteristics of HPLC, FPLC, and LC systems**

| System | HPLC | FPLC | LC |
| --- | --- | --- | --- |
| Support material (diameter, $\mu$) | Glass (5–10) | Polysaccharide* Cross-linked (50) | Polysaccharide (50–100) |
| Pressure | High | Intermediate | Low |
| Column dimensions | 5 × 50, 10–300 mm | 10–250 mm | 15–1000 |
| Sample size (separation type) | 5–50 μl (analytical) | 1 ml (analytical) | 10 ml (preparative) |
| Analysis time | 10–15 min | 30 min | 8–24 h |

HPLC, high-pressure liquid chromatography; FPLC, fast-protein liquid chromatography; LC, low-pressure chromatography.

chromatography column. A high-pressure pump is needed to propel the mobile phase. All flow-lines and the column material are made from steel in order that these components can withstand high pressures. Apart from such design features, there is no intrinsic difference between HPLC and low-pressure liquid chromatography. HPLC, FPLC (fast-protein liquid chromatography), and LC (liquid chromatography) employ increasingly larger diameter supports and lower pressures for separation. Characteristics of HPLC, FPLC, and LC systems are given in Table 2.4.

**Gas chromatography** uses gas as the mobile phase. The column consists of a narrow-bore glass tube coated with a nonpolar stationary phase. A common detector is a **flame ionization detector** (FID). This consists of two electrodes placed on either side of a flame. Material eluted from the GLC system is ionized in the detector flame leading to a flow of current between the electrodes. Some GLC systems are directly coupled with mass spectrometers. These so-called GC–MS systems are invaluable for the analysis of volatile species encountered in flavor research. With modern computerized systems information from the GC–MS is automatically

---

**Table 2.5**
**Types of column liquid chromatography separation**

Ion exchange chromatography (IEC)
Anion exchange using CM support
Cation exchange using DEAE support
Chromatofocussing
Hydrophobic interaction chromatography (HIC)
Size exclusion chromatography (SEC)
Affinity chromatography

CM, carboxymethyl, DEAE, diethylaminoethane.

---

compared with a huge database of stored information. It is possible to identify the compounds eluted from a chromatogram by matching their mass spectra with a library of standard results stored in the GC–MS instrument library.

## 2.3.2 Modes of Chromatographic Separation

Some of the common chromatography systems used by food scientists are listed in Table 2.5, and key components are illustrated in Figure 2.4. It is the column packing that is unique. The support material is either beaded glass or polysaccharide (agarose, Sephadex, Sepharose). The former material is more rigid, less compressible at high pressures and therefore more suited to HPLC applications. Whether polysaccharide or glass, the support material first undergoes chemical modification to increase its strength and/or place specific chemical functions on the support surface. All support categories (Table 2.5) can be used for HPLC, FPLC or LC. Details of some of support types and the mechanism(s) of sample separation are described below.

### 2.3.2.1 Ion Exchange Chromatography

In ion exchange chromatography the column material is covalently bonded with positive charges derived from DEAE (diethylaminoethylethane) or negative charges from CM (carboxymethyl) groups. Strongly anionic supports have surface sulfonate groups. The sample is injected into the column followed by a continuous stream of mobile phase to wash out unbound material. Material adsorbed to the column by ionic interactions is later eluted by injecting a mobile phase containing increasing concentrations of a salt. With **gradient elution**, the mobile phase contains a gradually increasing concentration of salt. Pre-manufactured DEAE and CM columns are available for HPLC. The technique has been successfully used to analyze polar compound including **organic acids** produced by fermentation, **carbohydrates** (simple sugars), **phytochemicals** from wine, **amino acids**, **peptides**, and **proteins**. The ability to separate and quantitatively determine a large number of sugars and metabolites from fermentation, using the Aminex$^{TM}$ column and a refractive index detector is particularly impressive.

### 2.3.2.2 Hydrophobic Interaction Chromatography

The support material for hydrophobic interaction chromatography (HIC) is either carbohydrate or glass beads to which are attached nonpolar groups. Typically, these are phenyl or straight-chain alkane compounds containing C8, C12 or C18 carbon atoms. With low-pressure separations, the sample to be separated is dissolved in a solvent with a high concentration of a salt such as ammonium sulfate, which enhances nonpolar interactions with the column. Sample components that are more hydrophobic adsorb to the column strongly, whereas polar components are eluted first. To separate adsorbed material, the HIC column may be washed with a mobile phase having a lower ionic strength. Increasing concentrations of polar additives like ethylene glycol or alcohol can also be added to the elution buffer to remove material bound strongly to the HIC column.

### 2.3.2.3   Reverse-Phase High-Pressure Liquid Chromatography

The stationary phase for RP-HPLC is usually glass beads modified with octadecyl (C18) groups. This technique has been successfully applied for chromatophic analysis of a large number of food micro-components, such as amino acids and **peptides, colors, lipids, phytochemicals**, and **vitamins**. Indeed, RP-HPLC is probably best considered the default technique for vitamin analysis. The mobile phase is generally a mixture of water and a water-miscible organic solvent such as acetone, acetonitrile or dimethylsulfoxide. A common method for peak detection during RP-HPLC analysis is to use an ultraviolet detector (set at 250 or 280 nm). With food components that do not absorb UV radiation, a refractive index (RI) detector may be used.

### 2.3.2.4   Size-Exclusion Chromatography

Size-exclusion chromatography (SEC) separates components on the basis of their size. The technique is also called **gel permeation chromatography** or **gel filtration analysis**. The column for SEC is filled porous beads. As a sample flows through the SEC column, low molecular weight compounds diffuse into the interior of the support beads. In contrast, larger molecules are increasingly excluded from the support phase. Those molecules having a larger hydrodynamic volume or Stokes radius are eluted from the SEC column first. Gel permeation chromatography is a form of filtration as high molecular weight compounds pass through the "filter" first.

Applications of SEC in food science occur in biopolymer (protein and polysaccharide) mass determination. Information obtained from starch analysis includes degree of polymerization and extent of branching. SEC can also be used for de-salting samples: as simple salt ions are much smaller than biopolymers they are retained to a greater extent by an SEC column.

### 2.3.3   THE EFFICIENCY OF CHROMATOGRAPHIC SEPARATIONS

The result of a chromatographic experiment is a **chromatogram** (Figure 2.5). A hypothetical chromatogram obtained by separating a two-component mixture shows two peaks (peaks 1 and 2). In each case the peak height or peak area is proportional to the concentration of the chemical species. However, the relative heights or areas under peaks 1 and 2 are not proportional to the relative concentrations of components 1 and 2. Put another way, injecting equal concentrations of each component by itself will not produce peaks of equal size. The response to each component in a mixture is determined by the detector sensitivity for that component. With a photometric detector, the instrument wavelength should be such that the components of interest absorb equally.

The procedure for assessing chromatographic performance is straightforward. We may wish to optimize the efficiency of a chromatographic separation. The number of variables can be manipulated in a bid to improve the degree of sample resolution including the choice of column support dimensions, column size, solvent composition, mobile phase flow rate, and the

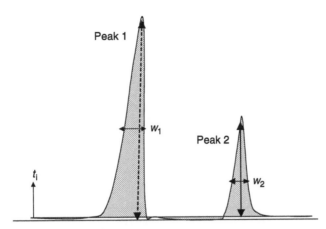

**Figure 2.5** A chromatogram showing the separation of two components. The sample was injected at the injection time ($t_i$). Peak heights and peak widths at half-peak heights are shown. The period from $t_i$ to any peak maximum is the retention time ($t_R$).

volume of sample injected. Sample resolution ($R$) can be calculated for any two peaks using the relation

$$R = \frac{2(t_{R2} - t_{R1})}{w_1 + w_2},$$  (2.4)

where $w_1$ and $w_2$ are peak widths at half the peak height and $t_{R1}$ or $t_{R2}$ are the retention times (time elapsed from sample injection to the peak maximum). A high resolution leads to wide separation between peaks (increasing values for $t_{R1}$ minus $t_{R2}$) and sharp narrow peaks (decreasing values for $w_1$ plus $w_2$).

The degree of sample peak broadening inside a column determines the separation efficiency measured as the height equivalent theoretical plate (HETP). The HETP is a notional partitioning event that contributes to overall sample separation. A chromatographic column with a large number of HETPs will produce a more complete separation of two components as compared to a column with a low number of HETPs. This is a function of several factors that influence sample spreading including the following:

(1) Peak broadening arising from sample molecules flowing around particulate packing material and in between gaps in the column packing. This eddy diffusion contribution to sample dispersion is independent of mobile phase flow rate.

(2) Diffusion of sample molecules into the porous packing material. This partitioning process increases with decreasing mobile phase flow rate which leaves more time for sample equilibration between the mobile and packing phase.

(3) Longitudinal diffusion of sample components. This form of sample spreading decreases with *increasing* solvent flow rate. Hence, longitudinal spreading will increase immensely as the mobile phase flow rate is reduced close to zero.

These ideas are summarized by the **Van Deemter equation**:

$$\text{HETP} = A + C\Phi + \frac{B}{\Phi}$$  (2.5)

where $\Phi$ represents the mobile phase flow rate. The terms $A$, $B$, and $C$ include such parameters as the support-packing diameter, diffusion coefficient for sample molecules, and

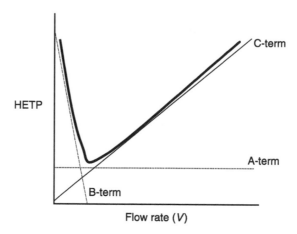

**Figure 2.6** A schematic presentation of how the three terms of the Van Deemter equation are affected by the mobile phase flow rate (V) during a chromatographic separation.

column dimensions. The effect of mobile phase flow rate on the HETP can be visualized as shown in Figure 2.6. Clearly, the first term ($A$) is not affected by solvent flow rate. Finally, it is worth noting that the HETP value can be determined from empirical measurements of sample peak width ($W$) and retention time ($t_R$):

$$HETP = \frac{\ell}{16}\left(\frac{w}{t_R}\right)^2.$$  (2.6)

In closing, empirical relations (2.4) and (2.6) allow us to evaluate the performance of a given chromatographic system. It is then also possible to consider how a range of experimental variables may be altered in order to optimize column performance.[4]

## 2.4  PHOTOMETRIC ANALYSIS

### 2.4.1  Introduction

Some analytical applications of visible radiation are based on visual measurements of color. The color of tomatoes or apples can be measured by comparison with a reference chart. The use of instruments for the measurement of color dates from about 1933 when Bauman and Steenbock measured vitamin A and carotene in butter. In 1938 Prill and Hammer described the colorimetric determination of **diacetyl** as the iron-diacetyl-glyoxime complex. The technique was applied to ice cream or butter.

Shining a light onto a surface leads to **absorption**, **reflection**, and **refraction**. In absorption colorimetry, we relate the quantity of transmitted light to the concentration of the absorbing species. Colorimetry involves measurements at a single wavelength. Spectrophotometry involves photometric measurements at multiple wavelengths.

### 2.4.2  Properties of Electromagnetic Radiation

Light is electromagnetic radiation. This consists of electric and magnetic fields oscillating at right angles to each other and to the direction of wave propagation. Some important characteristics of electromagnetic radiation are listed in Table 2.6.

| Table 2.6 Characteristics of electromagnetic radiation | |
|---|---|
| **Parameter** | **Dimensions** |
| Wavelength ($\lambda$) | meters (m) |
| Velocity ($c$) | meters per second (m/s) |
| Frequency ($v$) | per second (s) |
| Energy ($E$) | joules |

The energy ($E$) associated with electromagnetic radiation depends on its frequency. The relation between wavelength, frequency, and velocity of light ($c$) is also of interest:

$$E = hv, \tag{2.7}$$

$$c = v\lambda, \tag{2.8}$$

where $h$ ($6.63 \times 10^{-34}$ J s) is Planck's constant. The electromagnetic radiation spectrum is as follows: radio waves ($\lambda = 50$–$100$ cm), microwaves, infrared (far-IR, mid-IR, near-IR), visible, UV-light, vacuum UV, X-rays, gamma radiation.

### 2.4.3 ABSORPTION PHOTOMETRY

#### 2.4.3.1 Absorption of Electromagnetic Radiation

When electromagnetic radiation passes through a material it is affected by the local electric and magnetic fields in that material. These force fields are due to electrons orbiting around the nuclei of atoms or else forming chemical bonds. The incoming electromagnetic radiation produces one of three effects depending on the energy:

- Gentle vibration of groups—infrared radiation
- Excitation of electrons to high energy orbitals—visible and UV radiation
- Ionization—very high energy electromagnetic radiation

Successful transfer of energy requires a match between the wavelength or frequency of the incident radiation and the physical process. We may suppose that incident electromagnetic radiation is absorbed and re-emitted. A 100% re-emission causes no apparent effect. Less than 100% re-emission causes a reduction in the intensity of re-emitted light. Slow re-emission reduces the velocity of the incident radiation in proportion to the refractive index ($n$) of the material:

$$Refractive\ index\,(n) = \frac{velocity\ of\ light\ (sample)}{velocity\ of\ light\ (vacuum)}. \tag{2.9}$$

#### 2.4.3.2 Effect of Chemical Groups on Light Absorption

The absorption of vis/UV light is mainly due to pi electrons located within double bonds. Ethylene absorbs at a wavelength of 170 nm. The C=O bond of acetic acid absorbs at 208 nm.

The primary absorbing groups (Table 2.7) are **chromophores**. The structure of a chromophore affects $\lambda_{max}$. The presence of conjugated double bonds (or alternating single–double bonds) increases the wavelength of absorption. For instance $\lambda_{max}$ is 171 nm

**Table 2.7**

**Some chromophores and their characteristic absorbances**

| Chromophore | Chemical grouping | $\lambda_{max}$ (nm) | $\varepsilon_M$ ($M^{-1} cm^{-1}$) |
|---|---|---|---|
| Ethylene | C=C | 171 | 15,530 |
| Acetic acid | COOH | 208 | 23 |
| Diazo | N=N | 338 | 4 |
| Aldehyde/ketone | R.CO.R | 160–166 | Approx 20,000 |

$\lambda_{max}$ = wavelength of maximum absorption; $\varepsilon_M$ ($M^{-1} cm^{-1}$) = molar extinction coefficient.

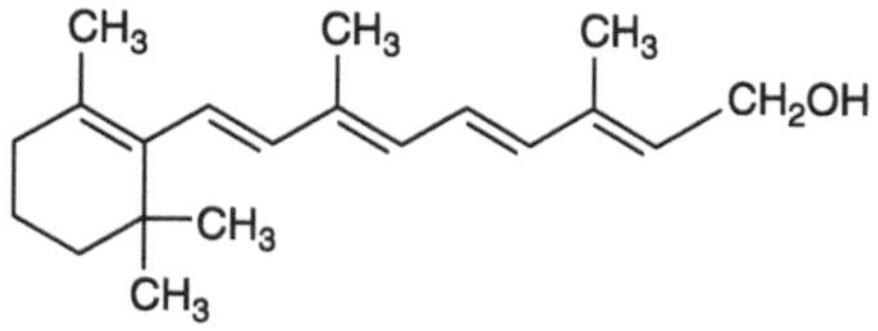

**Figure 2.7**   The structure of vitamin A ($\lambda_{max}$ = 360 nm).

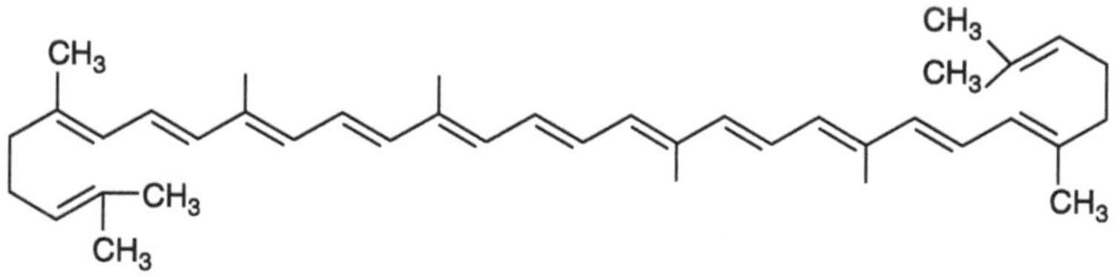

**Figure 2.8**   The structure of lycopene from tomato ($\lambda_{max}$ = 480 nm).

for ethylene compared with 360 nm for vitamin A (Figure 2.7) or 470 nm for lycopene (Figure 2.8) isolated from tomatoes.

$\lambda_{max}$ also depends on non-absorbing constituents (auxochromes). Electron donating species lower $\lambda_{max}$. For benzene $\lambda_{max}$ = 254 nm. Addition of a methyl or hydroxyl group to form toluene or phenol produces $\lambda_{max}$ values of 261 and 280 nm, respectively.

### 2.4.4   QUANTITATIVE ABSORBANCE STUDIES

#### 2.4.4.1   Beer and Lambert's Law

Light absorption is used for quantitative analysis. The absorption of incident light is proportional to the intensity of the incident light ($I$), sample thickness ($L$) and the concentration of dissolved solute ($C$). Equation 2.10 shows that light intensity decreases as a single wavelength of light passes through a sample. Slight rearrangement followed by integration gives Beer's law:

$$\frac{dI}{dL} = \varepsilon CI \quad \text{and} \quad \int \frac{dI}{I} = \varepsilon c \int dL, \tag{2.10}$$

$$ln\left(\frac{I_0}{I}\right) = A = \varepsilon CL, \tag{2.11}$$

$$A = \varepsilon CL. \tag{2.12}$$

The logarithmic term in Eq. 2.11 is the absorbance ($A$). According to Beer's law, a graph of absorbance plotted versus sample concentration yields a straight-line plot with a slope of $\varepsilon$ ($M^{-1}cm^{-1}$). $\varepsilon$ is defined as extinction coefficient. In practice, the linear range for a calibration graph is limited to absorbance values of $\leq 3$. Deviations from a linear relationship arise if the sample is highly concentrated. Under such circumstances chromophores interact by hydrogen bonding. The presence of particulate suspensions in the sample will also affect linearity.

### 2.4.4.2  Practical Aspects

Colorimetry, the measurement of sample absorbance, requires certain precautions in order to provide reliable results:

(1) Choice of cuvette. Measurements using UV wavelengths (<320 nm) require quartz cuvettes. For visible wavelength measurements glass cuvettes can be used. The former type is 10–100 times more expensive and usually marked with the letters Q or UV.

(2) Sample filtration. There should be no particulates. Avoid very high absorbance readings. If readings of over 2 are obtained, dilute the sample by a known amount and remeasure the absorbance.

(3) Blank readings. Measurements should be compared with the "reagent" blank. Design of a reagent blank is difficult. Ideally it should be identical to the sample except that it should not contain the analyte.

## References

1. Skoog, D. A. and West, D. M., *Fundamentals of Analytical Chemistry* (3rd Edition), Holt-Saunders, Philadelphia, PA, 1976.
2. Hart, F. L. and Fisher, H. J., *Modern Food Analysis*, Springer-Verlag, New York, 1–27, 1974.
3. Gruendwedel, D. W. and Whitaker, J. R., Eds, *Food Analysis*, Marcel Dekker, New York, 55–133, 1987.
4. Hart, F. L. and Fisher, H. J., *Modern Food Analysis*, Springer-Verlag, New York, 649, 1974.

# 3

# Statistical Analysis

## 3.1 INTRODUCTION

Statistical methods are essential for rational decision making and for scrutinizing data for their reliability.[1] Statistical analysis is necessary to determine whether one value is significantly different from another. Questions about the efficacy of various experimental treatments can also be answered. The mind-set and vocabulary introduced from statistics is essential for the practicing food chemist. For instance, the word **significant** has a specific meaning in the context of statistics. Therefore, statistics is a cornerstone of food chemistry. Table 3.1 shows a typical sequence of topic development in elementary statistics. The current treatment is intentionally steamlined. There are no mathematical proofs and we get from a to b with a minimum of fuss. Further information can be found in Petrie and Sabin[2] and Sokal and Rolf.[3]

Any measurement regardless of how it is performed, is subject to **uncertainty**. These errors cannot be eliminated because their origins are not known. It is for such reasons that measurements need to repeated. Replicated results are representative results. **Reproducibility**, the nearness of repeated observations to a central value, is an essential hallmark of scientific observation. Obtaining reproducible results is a major pre-occupation for analytical chemists. Statistical methods are necessary to deal with uncertainty inherent in all data.

### 3.1.1 Types of Data and Histograms

Quantitative data comes in two forms: **numerical data** and **categorical data** (yes/no responses)[a]. Numerical data is further divided into discrete (whole number values) and continuous data. One way of handling large amounts of data is to transform them into a pictorial form. An assortment of graphical displays are available. For instance, customer survey results can be presented as a **histogram** or **pie chart**. A calibration plot is usually a **line graph** showing the independent variable (controlled by the experimenter) plotted on the $x$-axis versus the dependent (measured) variable on the $y$-axis. Some rather advanced graphical displays involve 3-dimensional surface plots and **wire frame plots** etc.

In most laboratory experiments it is typical to make a minimum of two to five repeated measurements for each data point. The **average** value determined from these replicate observations is then adopted as the representative value. Actually, there are three kinds of averages (mean, median or mode) each with its own disadvantages and advantages. The discussion will be restricted to the mean. Variation or scatter obtained for a set of results can be assessed using several parameters including the variance ($s^2$), standard deviation ($s$), and coefficient of variation (CV).

---

[a]Some authors refer to categorical and numerical data as qualitative and quantitative, respectively.

## Table 3.1
## Important themes in statistics

| Item | Comment |
|---|---|
| 1. History of statistics, and types of data | Some historical details |
| 2. Pictorial display of data | Graphs, histograms, and trends |
| 3. Descriptive statistics — summarizing data, measuring averages and scatter | Describe the (a) mean, mode, median, and (b) sum of square, variance, standard deviation, or coefficient of variation |
| 4. Normal and other probability distributions e.g., $t$-, binomial, chi-squared distribution | Area segments under a bell-shaped graph show the degree of probability; the whole curve is probability of one |
| 5. Confidence statements | 95% chance says the true result (were it to be available) is like ours |
| 6. Testing hypotheses | There is a 95% chance that the new invention has no effect: the **null hypothesis** and the alternative |
| 7. Correlation and regression | Related to calibration graphs |

### 3.1.2 THE POTATO CHIP SCENARIO

Statisticians seem to see the world somewhat differently from other people. To get into the statistical mode of thinking, consider the following quality control problem. We wish to determine the weight of a packet of potato chips. The resulting information may be needed for labeling purposes. Such data can also be used to check the performance of our packaging process line. The weight of a potato chip packet can be determined by using a weighing scale but there is more to the potato chip scenario than meets the eye on first inspection.

First, there will be weight differences for any two packets of potato chips collected at random. The differences are due to human or equipment error and slight differences in how the production line works. In the worst case scenario, an earring or some machine fragment may have inadvertently entered the product stream and found its way into a package. The weights for packets of potato chips will also vary due to errors in measuring the weights. The chances of finding two packets with exactly the same weight are slim but not zero. The problem is how to obtain a representative weight for potato chips.

The simplest way to proceed would be to weigh every packet of potato chips produced in a single production run. We could then calculate the average weight and report this as the representative weight. We can object to this simple approach for two reasons: (1) this simple approach is highly wasteful of resources, including time and money; and (2) knowing the weight of every packet of potato chips does not allow *future* decision making.

According to statistical theory, a packet of potato chips has a true weight designated by the Greek symbol $\mu$. This true weight is a theoretical weight *because it is not accessible to direct measurement*. We cannot discover the value for $\mu$ — also called the population value — due to **indeterminate** error (Figure 3.1). Each single weight measurement ($Y_i$) or *sample* measurement is an estimate for the true weight, $\mu$. Moreover, if we take the weight for six packets of chips we can add them and divide by 6 to find the **mean** weight ($\hat{y}$). Since $\hat{y}$ is determined using only six packets of chips, the mean weight is only one estimate for the "true" weight $\mu$. To determine this true weight, we would need to weigh all packets of potato chips in our population. Deciding the nature of the "population" to which our results will ultimately be applied takes some consideration. In the potato chip scenario the population comprises all the packets of

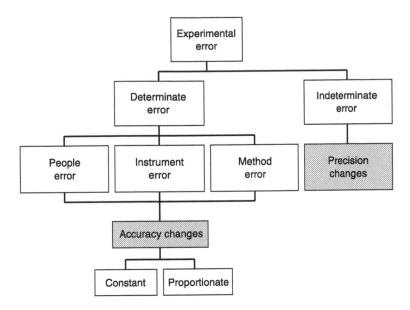

**Figure 3.1.** Types of error encountered in experimental science.

chips produced in a single production run. To obtain a really reliable estimate for $\mu$ we could also determine the average weights for several *different sextuplets of potato chip packets*. The average of six $\hat{y}$ values would then be more representative of $\mu$.

With increasing numbers ($n$) of weight measurements, the value for $\hat{y}$ approaches closer to $\mu$. In principle, the difference between the sample and population means will disappear ($\hat{y}-\mu=0$) if we make an infinite number of measurements. Statistical parameters such as $\hat{y}$ which are determined by experiments are called "sample statistics" in contrast to population statistics which are theoretical or true values.

To return to the potato chip scenario, it is time to formulate an efficient procedure for obtaining the required data. First, remove several sets of six packets from the production line at random and record their individual weights. Second, using Eq. 3.1 determine the average weight. Upon successful completion of this excercise, we should have obtained six different values for $\hat{y}$. Third, determine the average of six averages in order to find a representative weight for the potato chips. In the next section, we explain the rationale behind this three-step method of finding the average weight. The issues raised by the potato chip scenario are common for the quality control specialist. First, some important statistical terms are defined in Sections 3.2.1 and 3.2.2.

## 3.2  DESCRIPTIVE STATISTICS

### 3.2.1  THE MEAN OR AVERAGE

The mean ($\hat{y}$) is the sum of all observations divided by the number of measurements ($n$). The symbol $\Sigma$ reads "take the sum of." $Y_i$ refers to the weights of individual potato chip packets. By weighing six packets individually we obtain six results ($y_1, y_2, y_3, y_4, y_5$, and $y_i$). Notice that the $i$ subscript in Eq. 3.1 counts the number of data from 1 to 6. As the number data points increase to very large values, the mean approaches a theoretical value called the population mean ($\mu$).

$$\hat{y} = \sum (Y_i)/n \tag{3.1}$$

The mean ($\hat{y}$) is sometimes called a summary statistic[b]. A summary statistic summarizes information contained in the original (six) observations. Another type of "average" is the **median**. This is a data value that has an equal number of smaller and higher readings. The **mode** is the most frequent observation.

### 3.2.2   Measures of Error and Scatter

Absolute error is the difference between $Y_i$ and the accepted value ($Y_{true}$). When expressed as a relative quantity we get the relative error (Eq. 3.2). We can measure the degree of scatter for multiple results using the **variance** ($s^2$) or **sum of squares** (SS) expressed by Eqs 3.3 and 3.4. The SS is useful for some advanced forms statistical analysis such as analysis of variance (ANOVA) described later.

$$\text{Relative error} = (Y_i - Y_{true})/Y_{true}. \tag{3.2}$$

$$\text{Variance} (s^2) = \sum (Y_i - \hat{y})^2/(n-1). \tag{3.3}$$

$$\text{Sum of squares} = \sum (Y_i - \hat{y})^2. \tag{3.4}$$

The square root of the variance called the **standard deviation** (SD) which is another useful measure of scatter. It is good practice to report the value for the mean results plus/minus the standard deviation ($\hat{y} \pm$ SD). We should note also that SD is an estimate of the population standard devation ($\sigma$). The value of SD approaches $\sigma$ as the number of data points increases. Values for SD and its theoretical cousin $\sigma$ have the same dimensions as the initial measurement ($Y_i$). If a packet of chips is weighed in grams, the mean and SD will have dimensions of grams. It is sometimes convenient to express error (SD) as a percentage of the mean. The coefficient of variation (CV) is obtained by dividing SD by the sample mean:

$$CV = (SD/\hat{y}) \times 100. \tag{3.5}$$

The CV, being a ratio, has no units and is described as *dimensionless*. CV is useful for comparing the degree of scatter for measurements with greatly varying dimensions. The scatter for tree height measurements (in feet) can be compared with scatter for potato chip data when both types of results are converted into CV (%). The variance, sum of squares, standard deviation, *CV*, and standard error of the mean are all measures of **precision**.

### 3.2.3   Standard Error of the Mean

Whilst SD is often presented with experimental data, it is the **standard error of the mean** (SEM) which is of interest. The expression converting SD to SEM shown in Eq. 3.6. The SEM is the degree of scatter associated with a *collection of means*. This notion takes some thinking about. Recall from the potato chip experiment that six sets of $\hat{y}$ values were recorded. Each of these means is an estimate of the population mean, $\mu$. To obtain a still more reliable estimate for the population mean, we should find the mean of the means. The standard deviation of means is

---

[b]Example: Find the mean for the five observations (4, 5, 6, 7, and 14). Answer: $\Sigma(Y_i) = 36$. From Eq. 3.1 divide 36 by 5 which is $\hat{y} = 7.2$.

the SEM, a parameter that *is used in virtually all forms of statistically based decision making*:

$$SEM = SD/n^{1/2}. \tag{3.6}$$

### 3.2.4   TYPES OF STATISTICAL ERROR

Experimental error is of two kinds (Figure 3.1). **Determinate error** is traceable error due to people error, instrument error or method error. **People error** (sometimes called operator error) is due to poor technique, carelessness or momentary lapses in concentration. **Instrumental error** arises from poor machine maintenance or adjustment. **Method error** is due to inherent limitations in the experimental method. For instance, the titration indicator reactions might be slow or incomplete. Old or partially degraded reagents and the presence of interferences are other possible sources of method error. Determinate error reduces accuracy[c]. People error can be reduced by training and improved motivation. Instrument and method errors can be corrected by perfoming one of several types of calibration (see later). Another type of uncertainty is called **indeterminate error**. These are numerous small errors of uncertain origin and they cannot be eliminated entirely.

## 3.3   DISTRIBUTION CURVES

### 3.3.1   NORMAL DISTRIBUTION CURVE

Recalling the potato chip scenario, a graph showing the number of packets (y-axis) versus weights (x-axis) would follow a bell-shaped curve called the normal distribution[d]. An example of a normal curve is shown in Figure 3.2. For an infinite number of measurements the most frequent weight recorded would coincide with the theoretical mean, $\mu$. A small number of potato chip packets would weigh more than $\mu$. A small number of packets will possess weights far below $\mu$. The equation for the normal distribution (Eq. 3.7) *looks* formidable:

$$frequency = \frac{1}{\sigma\sqrt{2\Pi}}\exp\left(-\frac{(Y-\mu)^2}{2\sigma}\right). \tag{3.7}$$

The shape of a normal distribution curve is controlled by two parameters, $\mu$ and $\sigma$. The value for $\mu$ positions the curve on the x-axis whilst $\sigma$ determines the degree of spread. The total area under a normal distribution curve shows all the theoretical results possible. *The total area under a normal distribution curve represents a probability of 1*. From theoretical calculations, the region from $z=1$ to $z=-1$ contains 68% of the area under a normal curve. The area bounded by $z=-2$ to $z=+2$ contains 95% of the area under the normal curve. The interval bounded by $z=-3$ to $z=+3$ of a normal curve contains 99.5% of the area under a normal curve.

---

[c]Accuracy is defined as the closeness of a series measurement to a true value. By contrast, precision is a measure of the scatter of results measured in relation to the mean.
[d]To use the normal distribution curve for different types of studies, values on the x-axis are standardized by converting to a new index, Z. This operation involves subtracting the population mean from each measurement $(Y-\mu)$ and dividing by the standard deviation.

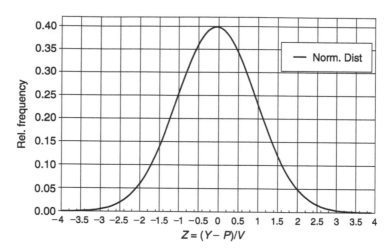

**Figure 3.2.** The normal distribution curve is bell-shaped.

**Figure 3.3.** The inventor of the *t*-distribution **William Gosset** (1876–1937) worked as a (food) chemist in the Guinness brewery in Dublin in 1899. He invented the *t*-test to handle small samples for quality control in brewing. He wrote under the name "Student" so the *t*-test is more commonly called Student's *t*-test. (From http://www-gap.dcs.st-and.ac.uk/~history/PictDisplay/Gosset.html.)

Assuming values for $\sigma$ and $\mu$ could be found, we would predict that 68% of all potato chip packages would weigh $\mu$ ($\pm\sigma$) grams and that 95% of all packages would weigh $\mu$ ($\pm 2\sigma$) grams and a chip packet selected at random would have a 95% chance of weighing $\mu \pm 2\sigma$ grams. The area shown under a normal distribution curve presents the probability that our results will fall within certain ranges of values defined in terms of $\mu$ and $\sigma$. *It is really important to understand this notion before proceeding.* A further important point is that since values for $\mu$ and $\sigma$ cannot be measured the theoretical normal distribution curve has no practical use. Recall from Section 3.2 that population statistics such as for $\sigma$ and $\mu$ are not directly measurable.

### 3.3.2 *T*-Distribution

There is another distribution curve called the *t*-distribution, which is applicable to real-life problems (Figure 3.3). The *t*-distribution shares many of the characteristics of the normal distribution. However, the former can be used when the number of observations is small. For the *t*-distribution the values for *Z* (see *x*-axis in Figure 3.2) are replaced by the *t*-parameter defined by Eq. 3.8:

$$t = (Y_i - \hat{y})/s, \tag{3.8}$$

where $Y_i$ is the value for each "individual" potato chip packet, *s* is the sample variance, and $\hat{y}$ is the sample mean. Unlike the normal distribution curve, the *t*-distribution is based on measurable parameters ($\hat{y}$ and *s*). Using the *t*-distribution, we may state that the intervals $\hat{y} \pm 1s$, $\hat{y} \pm 2s$, and $\hat{y} \pm 3s$ contain 68.3%, 95%, and 99% of all results. There is a 95% chance of finding a "true" result within an interval $\hat{y} \pm 2s$. These rather impressive predictions are based on *sample* estimates of the mean and standard deviation as described in Eqs 3.1 and 3.3. These calculations can be easily determined using a hand-held calculator. This is an important advance, because our potato chip experiment had only three to six measurements and we could not directly access the population statistics $\mu$ or $\sigma$.

### 3.3.3 CONFIDENCE INTERVALS

It will be interesting to estimate, with some confidence that the "true" weight of a packet of potato chips lies within a certain range of values. To proceed, we need *actual* values for the mean and standard deviation for packets of potato chips. The formal method for calculating confidence intervals is described in Table 3.2. The crux of the method is the SEM as described in Section 3.2.3.

A *t*-table is a list of values for *t*-parameters and their corresponding proportion of area under the *t*-distribution curve. For example, 95% of the area under a *t*-distribution curve (assuming a reasonably large sample size) occurs between the interval $t_\alpha = -2.1$ and $+2.1$. The listed *t*-values increase in size with the degree of freedom (df). A *t*-table has degrees of freedom on the right-most column and values for $\alpha$ ($= 0.05$) across the top. The body of the table contains the values being sought.

To summarize, uncertainty prevents from us from ever measuring the "true" value of any physical object. By adopting statistical principles we can embrace uncertainty. The fundamental quantities needed are the sample mean and SD ($\hat{y}$ and *s*). Using these parameters we can obtain some impressive estimates of the true value or state with some degree of confidence that the "true" value will lie within a certain range of values.

---

**Table 3.2**
**Steps for calculating confidence intervals**

1. Required data:
    - mean ($\hat{y}$), number of measurements (*n*), standard deviation (*s*)
2. Calculate SEM
    - $s/n^{1/2}$ and find the degree of freedom (df) $= n-1$
3. For a confidence levels of 95%
    - $\alpha = (100-95)/100 = 0.05$
4. Go to a standard *t*-table and *read* the value for $t_{0.05}$
5. Calculate the 95% confidence interval
    - $\hat{y} \pm (t_{0.05} \times \text{SEM})$

---

## 3.4   TESTING A HYPOTHESIS

### 3.4.1   Testing a Hypothesis about a Mean

We often need to test a null hypothesis ($H_0$) that one result is not significantly different from some pre-set value versus the alternative ($H_1$) that it is. The following passage describes how this test may be performed.[4]

(1) Formulate the null hypothesis versus the alternative

- $H_0$: $\hat{y} = \mu$,
- $H_1$: $\hat{y} \neq \mu$.

(2) Decide on the probability level, e.g., 95% and look up $t_\alpha$ from a $t$-table with an appropriate degree of freedom. This was also done in the confidence level calculations described above.

(3) Assemble your data and calculate the mean and standard deviation.

(4) Calculate the $t$-parameter: $t = (\hat{y} - \mu)/\text{SEM}$.

(5) Compare the calculated $t$-value with $t_\alpha$ from the $t$-distribution table.

(6) Accept the $H_0$ provided $-t_\alpha < t < t_\alpha$, otherwise reject $H_0$.

There are other statistical tests for determining confidence levels or testing hypotheses. These differ in how to calculate $\hat{y}$ and SEM but they converge at point 4, above.

### 3.4.2   Testing the Hypothesis about Two Means

The null hypothesis is that two results are the same. Such tests are usually about sample means owing to the requirement for representative parameters. Recall that a single observation has no place within a scientific study. The problem is therefore how to test the null hypothesis, that two means are not significantly different versus the alternative hypothesis that they are different. The steps needed for a test about two means are outlined as follows. First, formulate the null hypothesis versus the alternative:

- $H_0$: $\hat{y}_1 - \hat{y}_2 = 0$,
- $H_1$: $\hat{y}_1 - \hat{y}_2 \neq 0$.

Recall that the SEM is an important parameter that measures the degree of error associated with a mean. When dealing with two results (means $\hat{y}_1$ and $\hat{y}_2$) we need to consider the total error associated with both. First, the sample variances are combined to generate the weighted average variance ($s_p^2$) as shown in Eq. 3.9. Next, the SEM for two samples is determined from Eq. 3.10. Finally, we determine the $t$-statistic from Eq. 3.11.

$$s_p^2 = \frac{(n_1 - 1)s_1^2 + (n_2 - 1)s_2^2}{(n_1 + n_2) - 1}. \tag{3.9}$$

$$\text{SEM} = \sqrt{s_p^2\left(\frac{1}{n_1} + \frac{1}{n_2}\right)}. \tag{3.10}$$

$$t = \hat{y}_1 - \hat{y}_2/\text{SEM}. \tag{3.11}$$

The conditions for accepting or rejecting the null versus the alternative hypothesis are as described in the previous section. Compare the $t$ and $t_\alpha$ as before. $H_0$ can be accepted where $-t_\alpha < t < t_\alpha$. Otherwise we reject $H_0$.

## 3.5   STATISTICAL PROCEDURES FOR CALIBRATION

In this section we describe how to apply statistical principles to a fairly common problem in food analysis—the construction of a calibration graph. Analytical methods need calibrating in order to ensure accuracy. Calibration procedures are also used as a means of detecting error. Common calibration procedures include (1) **method calibration**, which uses independent parallel methods of analysis; and (2) **analyses of standard samples**, including the use of reagent blanks and varying sample sizes.

### 3.5.1   METHOD CALIBRATION

Using independent or parallel methods of analysis a set of sample standards with known composition are analyzed using a new test method and a **reference method** that has been validated. A calibration graph is then drawn by plotting results from the reference method ($y$-axis) and the test results ($x$-axis). The $x_i$ and $y_i$ observations are related by an equation for a straight line:

$$y_i = ax_i + b, \tag{3.12}$$

where $a$ is the gradient and $b$ is the intercept for the calibration graph. For each $x_i$ result we can determine the *calculated* value ($y_{calc}$) via Eq. 3.13:

$$y_{calc} = ax_i + b. \tag{3.13}$$

By comparing values for $y_i$ and $y_{calc}$ we may determine the reliability of the new test method. For instance, the relative error or variance associated with a new test method can be assessed using pairs of $y_i$ and $y_{calc}$ values in the numerator of Eqs 3.2 and 3.3.

### 3.5.2   CALIBRATION USING STANDARD SAMPLES

The assay technique is assumed to be valid. We use standard samples to develop a calibration graph where the instrument response is plotted against concentration. By analyzing the reagent blank we can establish what proportion of our analytical response is due to the item being measured and what is due to interferences and background reactions. In Eq. 3.12, $x_i$ now represents known concentrations and $y_i$ are the corresponding instrument responses. Calibration factors ($a$, $b$, etc.) can be determined from simple algebra or statistical analysis of paired ($x_i, y_i$) results.

## 3.6   ANALYSIS OF VARIANCE (ANOVA)

In Section 3.4 we described the $t$-test for comparing two means. To perform a similar comparison of three or more means we use ANOVA.[4] The most common scenarios for using the ANOVA test are as follows. (1) Tests about means. In this instance, ANOVA is used to test the null hypothesis that all the means are the same versus the alternative hypothesis that one or more of the means are different. (2) Partitioning variance. Consider a two-stage experiment involving $k$ repeated samplings from the same batch of food and $r$ repeated analyses of each sample. ANOVA allows the separation of total error so that the contributions from the sampling step and the analysis part of the experiment can be determined.

### 3.6.1   ANOVA TEST FOR DIFFERENCES BETWEEN MEANS

In accordance with scenario 1, the aim of ANOVA is to establish whether a number of means from several groups of data are statistically different or not. The test is performed by assessing the total scatter within all our data and comparing this with the degree of scatter *between* groups. ANOVA makes use of the *sum of squares* (Eq. 3.4). A typical ANOVA calculation has three steps:

- Determine the total variation for all the data regardless of their group.
- Determine the degree of variation between groups, starting with the group means ($\hat{y}$); this step assumes automatically that group *means are the same*.
- Use an *F*-test to compare the total and between group variances.

The null hypothesis, which is that the total *versus* between group variances are equal, holds only when the means from which the between group variance is estimated are from the same population. Table 3.3 gives sample data and quantities need for the ANOVA calculation.

First, it is necessary to determine the **total variance** (mst) according to Eq. 3.9. In practice, it is more convenient to determine the total sum of squares (SST). Dividing the SST by the total degrees of freedom yields mst. Next, it is necessary to determine the **between group variance** (msc). This may be done by **treating** the means from column A, B, and C as initial data and using Eq. 3.3 to determine the msc. Table 3.4 shows a convenient formula for the msc. First, determine the between group sum of squares (SSC). Dividing SSC by the between

**Table 3.3**
**Example calculations for ANOVA**

|  | Treatment | | |  |
|---|---|---|---|---|
|  | A | B | C |  |
|  | 0.720 | 0.730 | 0.740 |  |
|  | 0.730 | 0.740 | 0.750 |  |
|  | 0.720 | 0.750 | 0.740 |  |
|  | 0.710 | 0.730 | 0.760 |  |
| $T$ | 2.880 | 2.950 | 2.990 | $\sum T = 8.820$ |
| $T^2$ | 8.294 | 8.703 | 8.940 | $\sum T^2 = 25.937$ |
|  |  |  |  | $\sum (x^2) = 6.4850$ |
| Mean ($\hat{y}$) | 0.720 | 0.738 | 0.748 |  |

**Table 3.4**
**Summary formulae for ANOVA**

| Sources of variation | Sum of squares | df | Variance | F-test |
|---|---|---|---|---|
| Total (SST) | $\sum x^2 - (\sum T)^2/rk$ | $rk - 1$ | $mst = \dfrac{SST}{rk - 1}$ | $\dfrac{msc}{mst}$ |
| Between groups (SSC) | $\sum T^2/r - (\sum T)^2/rk$ | $k - 1$ | $msc = \dfrac{SSC}{k - 1}$ | $\dfrac{msc}{msp}$ |
| Within a group (SSE) | $SST - SSC$ | $k(r - 1)$ | $msp = \dfrac{SSE}{k(r - 1)}$ |  |

**Table 3.5**

**Sample results for ANOVA. The table gives a worked example using data from Tables 3.3 and 3.4**

| Sources of variation | Sum of squares | df | Variance | F-test |
|---|---|---|---|---|
| Total (SST) | 0.0023 | 11 | 0.00021 | 3.58 |
| Between groups (SSC) | 0.0015 | 2 | 0.00075 | 8.52 |
| Within a group (SSE) | 0.0008 | 9 | 0.000088 | |

groups degree of freedom yields msc. Finally, the $F$-test is performed from the ratio of the between group and total variance.

Recall that our aim is to test whether data (means) obtained from three experiments, treatments, etc. (Table 3.3) are the same. In the final stage of the ANOVA we perform an $F$-test as follows:

$$F_{k-1,\,kr-1}\,\frac{msc}{mst} = \frac{0.00075}{0.00021} = 3.58. \tag{3.14}$$

In an $F$-table the results for a two-tailed test at the 5% significance level shows the $F_{2,11}$ critical value to be 5.256. Since the calculated $F$-value is smaller than the critical value we accept the null hypothesis (mean 1 = mean 2 = mean 3) and reject the alternative hypothesis that one or more of the means is significantly different. Comparing Tables 3.4 and 3.5 allows the reader to follow the preceding calculations.

### 3.6.2   ANOVA FOR PARTITIONING VARIANCES

Analytical techniques usually undergo collaborative testing. A group of laboratories each analyze a well defined sample using a pre-agreed method. The purpose of this sort of testing is to see whether a given technique performs acceptably at different laboratories. Alternatively, one or more laboratories may perform inadequately. The total precision for the test is the sum of error obtained within each laboratory and between laboratories. Large between laboratory variances could indicate that one laboratory did not perform the test properly or that the method under test needs improving. ANOVA can be used to separate the error arising within each laboratory from errors between laboratories. The situation will be illustrated using data from Table 3.3. Columns A, B, and C are hypothetical data reported from three laboratories.

The null hypothesis is that the within laboratory variance is no different from the between laboratory variance. The necessary calculations are summarized in Table 3.5.

$$F_{k-1,kr-1}\,\frac{between\ group\ VAR}{within\ group\ VAR} = \frac{msc}{msp} = \frac{0.00075}{0.000088} = 8.52,$$

$F_{2,9}$ critical value (two-tailed test, 5% significance) $= 5.715$. $\qquad\qquad$ (3.15)

Since the calculated $F$-value is greater than the critical value we reject the null hypothesis and accept the alternative that the between laboratory variance is significantly different to the within laboratory variance. Apparently, there are some differences in the way different laboratories perform the test. We might search for ways for reducing the between laboratory variation in results.

## References

1. Skoog, D. A. and West, D. M., *Fundamentals of Analytical Chemistry* (3rd edition), Holt-Saunders, Philadelphia, 42–88, 1976.
2. Petrie, A. and Sabin, C., *Medical Statistics at a Glance*, Blackwell Science Ltd, Malden, MA, 2000.
3. Sokal, R. R. and Rohlf, F. J., *Introduction to Biostatistics*, W. H. Freeman & Company, San Francisco, 1973.
4. Hayslett, H. T., *Statistics Made Simple*, W. H. Allen, London, 1976.

# 4

# Carbohydrates

## 4.1 INTRODUCTION

Table sugar (sucrose) and milk sugar (lactose) are the most well known simple carbohydrates. Starch and pectin are commonly encountered polysaccharides. Less well known are the plant gums including **gum arabic**, taracanth gum, and acacia gum. Bacteria exo-polysaccharides including **xanthan** are important as thickening agents in sauces and French dressing. The marine polysaccharides **alginate** and **carrageenan**, obtained from seaweed, are found in a range of products including substitute fruits and candy. **Chitin**, found in the exoskeleton of crabs, is beginning to find applications in the food industry. Humans and non-ruminants lack digestive enzymes for breaking down **cellulose, xylanose**, and **plant cell wall material**, which therefore function as **dietary fiber**. Most polysaccharides are **biodegradable** by bacteria in the environment. This chapter describes the properties of the carbohydrates and their role as food ingredients.

## 4.2 MONOSACCHARIDES

A carbohydrate is any organic compound with the formula $(CH_2O)_n$ with $n \geq 3$. Simple sugars having $n$-values of 3, 4, 5 or 6 are called **trioses, tetroses, pentoses** or **hexoses**. Monosaccharides are sugars that cannot be broken down by hydrolysis into other simpler sugars. Examples include glucose and fructose. Joining together two, three, a few or many monosaccharides generates a **disaccharide, trisaccharide, oligosaccharide** or **polysaccharide**, respectively. The interrelationships between different carbohydrates are shown in Table 4.1. Monosaccharides are the building blocks from which all other carbohydrates are made. Fundamental work on the nature of sugars was carried out by the German chemist Herman Fischer (Figure 4.1) during the nineteenth century.

### 4.2.1 MONOSACCHARIDES AND DISACCHARIDES

Simple sugars include monosaccharides such as glucose (dextrose), fructose (levulose), mannose, and galactose. Disaccharides such as sucrose (table sugar), lactose (milk sugar), and maltose (hydrolyzed starch) are also examples of simple sugars. Sucrose, fructose, and glucose are found in fruits and honey as natural sweeteners. Glucose is the most abundant sugar in the biosphere, being the major constituent of natural cellulose and starch. The chemical reactions of glucose provide a blue-print for the reactions of all other monosaccharides and more complex carbohydrates. In this section we will review the structure and reactions of glucose.

**Table 4.1**

**The interrelationships between different carbohydrates**

| Carbohydrates | Typical properties |
|---|---|
| Monosaccharide -●-, e.g., glucose | Sweet, soluble, crystallizable, metabolized, fermented |
| Disaccharide ●-●-, e.g., sucrose, lactose, maltose, cellobiose | Sweet, soluble, crystallizable |
| Oligosaccharide ●-●-●-●-●-●- | Dextrin |
| Polysaccharide (-●-●-●-●-●-●-)$_n$, storage (starch), structural (pectin, hemicellulose, cellulose), plant exudates, marine (alginate, carrageen), microbial (xanthan) | Mainly insoluble in cold, good gelling and thickening agents, not sweet, not crystallizable |

**Figure 4.1**   Herman Emil Fischer (1852–1919), German organic chemist famed for his work on sugars, purines, and peptides. Nobel Prize winner for chemistry 1901–1902.

### 4.2.2  GLUCOSE RING STRUCTURE

The general formula for glucose is $C_6(H_2O)_6$ showing glucose to be a hexose. The straight chain formula for glucose shows it to be a **polyhydroxyaldehyde**. This is any compound that contains multiple hydroxyl groups as well as an aldehyde group. Sugars that possess an aldehyde function are also called **aldoses**. The ketoses have a ketone group at C2 (e.g., fructose) are classed as a ketoses. The carbon atoms of glucose are generally numbered from 1 to 6 starting with the aldehyde at C1, which is called the **anomeric carbon**.

The **absolute configuration** of monosaccharides is determined by the disposition of the penultimate (C5) hydroxyl group. According to the glyceraldehyde convention D-sugars have an absolute configuration with the C5 alcohol group pointing to the right. On the other hand L-sugars have the C5 alcohol pointing to the left. Most sugars found in nature are D-sugars. The straight chain formula for glucose shows four **asymmetric carbon atoms**, each of which is attached to a different grouping. According to the **Van't Hoff rule**, this leads to $2^4$ (=16) possible hexose stereoisomers. The Fischer structure of some aldoses is shown in Figure 4.2.

The aldehyde and alcohol groups from the same glucose molecule react to form to a ring[a]. The intra-molecular reaction between the C1 aldehyde and C5 OH group produces a six-sided

---

[a]The C1 aldehyde group of glucose reacts with two moles of alcohol form a hemiacetal followed by a full acetal. The addition reaction is acid catalyzed. A glucose ring structure is an example of internal hemiacetal formation.

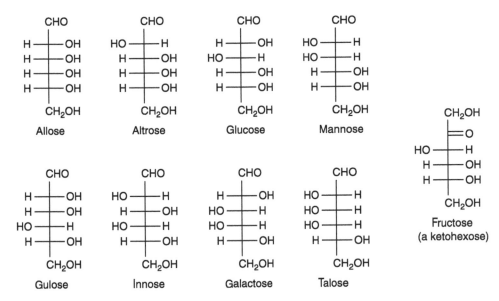

**Figure 4.2** The Fischer structures of some aldoses. These are all D-sugars.

**pyranose** ring. By contrast the reaction between the C1 (aldehyde) and the C4 (OH) group forms a five-member **furanose** ring. The terms furanose and pyranose are based on the names of two heterocyclic compounds, pyran and furan, which possess 6- and 5-membered ring structures, respectively. Internal hemiacetal formation leads to a newly formed OH group at C1 which points below the plane of the ring structure ($\alpha$-D-glucose) or above the plane of the ring ($\beta$-D-glucose) (Figure 4.3). The alpha and beta forms are additional isomers, or **anomers**, produced by ring formation by C1 which is therefore termed the anomeric carbon.

Figure 4.3 shows a number of different representations of the glucose ring structure. The **Hayworth** formula for $\alpha$-D-glucose shows the C1 OH group below the plane of the pyranose ring viewed sideways (structure III in Figure 4.3). This can be distinguished from the Hayworth structure for $\beta$-D-glucose (not shown) which has the C1 alcohol pointing *above* the plane of the pyranose ring viewed sideways. The **chair structures** for sugars (structure II, Figure 4.3) are similar to conformations formed by cyclohexane. The $^4C_1$ chair structure for D-glucose has C4 and C1 elevated and depressed relative to the plane of the ring. The structural equilibrium between the $^4C_1$ chair, boat ($^4C^1$), and $^1C_4$ chair structures favors the first structure: most sugars exist as the $^4C_1$ conformation.

The C1 OH group for $\beta$-D-glucose is directed at the **equatorial** plane for the chair structure. For alpha sugars the chair structure has an **axial**[b] hydroxyl group at C1. Starch is built from $\alpha$-D-glucose glucose units possessing axial C1 OH groups. Cellulose is a polymer of $\beta$-D-glucose units possessing equatorial C1 OH groups. The two different ($\alpha/\beta$) anomeric conformations for D-glucose account for the differences in the structure and characteristics of cellulose and amylose starch (Chapter 7, Section 7.5.3.2).

### 4.2.3 Reactions of Monosaccharides

The chemistry of carbohydrates is determined by two functional groups: alcohols and aldehydes. For example, alcohols undergo **esterification** with acids. Treatment with sulfuric acid,

---

[b]The terms equatorial and axial correspond loosely to latitude (side-to-side) and longitudinal (up-and-down) directions of the globe. The anomeric carbon–carbon atoms in Figure 4.3 (I) show equatorial and axial hydroxyl groups.

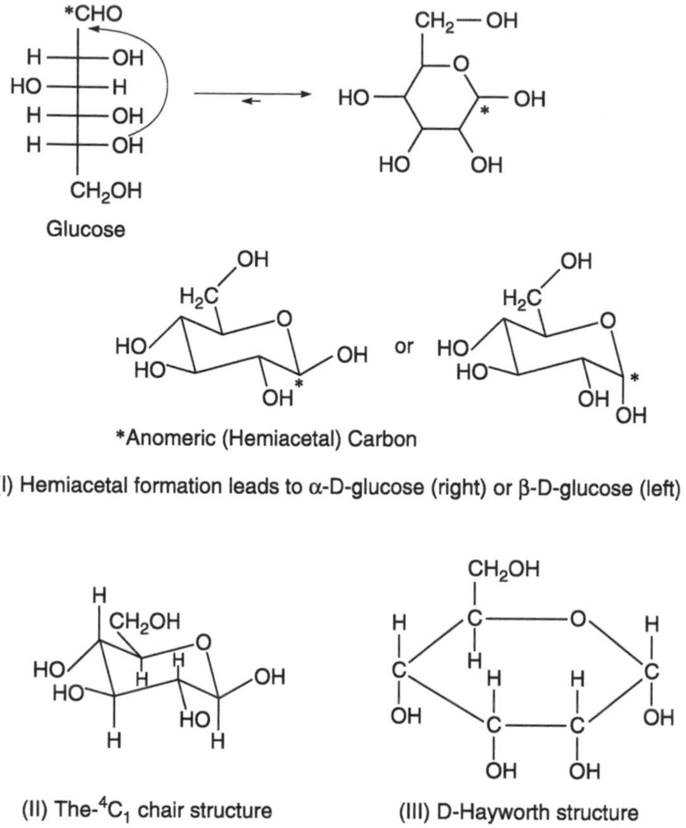

**Figure 4.3** Glucose ring structure. (I) Formation of a ring structure by internal hemiacetal formation. Two ways of drawing the glucose ring are shown. (II) The $^4C_1$ *chair structure for* β-D-glucose. The Hayworth presentation α-D-glucose.

---

**Table 4.2**

**Functional group chemistry of glucose as a polyhydroxyaldehyde**

| Alcohol group (–OH) | Aldehyde group (–CHO) |
| --- | --- |
| Esterification | Oxidation and reduction |
| Methylation | Hemiacetal formation |
| Dehydration | Addition with amines or hydrazines |

---

phosphoric acid or acetic anhydride forms glucose **pentasulfate**, **pentaphosphate** or **pentaacetate**, respectively. These same reactions can be applied to glucose polymers such as starch and cellulose. Upon treating with ethylene ethoxide the alcohol groups of glucose undergo **methylation** leading to an ether (R–O–Me). Methylation can also be achieved by reacting glucose with methyl iodide ($CH_3I$) under alkaline conditions. Methylation is a useful reaction for determining the ring size for sugars and the type of linkages between different monosaccharides. Other important reactions attributed to alcohol functional groups include **dehydration**. The functional group chemistry of glucose is summarized in Table 4.2. All these reactions can be performed with other monosacchrides.

The C1 aldehyde functional group in glucose behaves like other aldehdydes and it can easily be **oxidized**. Treatment with a mild oxidizing agent such as bromine water transforms the

CHO into a carboxylic acid (–COOH) group leading to D-gluconic acid. This is an important acidifying agent[c]. Aldehydes undergo **addition** reaction with amines (–NH$_2$) forming a **Schiffs base**. In many foods the –NH$_2$ group is located at the N-terminus of a protein chain or polypeptide. The epsilon NH$_2$ group from the amino acid **lysine** and alpha NH$_2$ group in other amino acids also react with sugars. Schiffs base formation is the first step in the Maillard reaction—also called nonenzymic browning (Chapter 10).

Sugars react with **phenyl hydrazine** via an addition process to form yellow products called **osazones** which crystallize from cold solution. Osazones formed by different sugars have different melting points and can be used to identify the parent sugar. The crystallization times for osazones from a hot solution also depend on the parent sugar: 0.5 min (mannose), 2 min (fructose), 4–5 min (glucose), 15–19 min (galactose), and 30 min (sucrose hydrolysis products). The osazones of maltose and lactose are soluble in hot water. Osazone formation is the result of **nucleophilic addition** between hydrazine and the anomeric carbon atom of an aldose (or C2 of a ketose). This is then followed by oxidation of C2 (or C1) of an alcohol group to a ketone (or an aldehyde). The new carbonyl group is also converted to a hydrazone. The reaction does not proceed further beyond C3. Finally, an aldehyde can be **reduced** to form an alcohol by treating with sodium borohydride. Glucose and other simple carbohydrates are transformed into polyhydroxy alcohols by reduction as described in the next section.

### 4.2.4  POLYOLS OR SUGAR ALCOHOLS

The C1 aldehyde group of simple sugars can be **hydrogenated**. The products of sugar hydrogenation are polyhydroxy alcohols or polyols (Table 4.3). Some polyols occur naturally in berries, apples, and plums. Industrially produced polyols are useful ingredients for the manufacture of hard candies, chewing gum, chocolates, baked goods, and ice cream.

The benefits suggested for sugar alcohols compared to normal sugars include reduced calorie content (2.1 compared to 4.0 kcal/g) which makes them useful for low calorie, sugar free foods. Sugar alcohols are being used to manufacture food products aimed at diabetics. These products are slowly absorbed from the digestive tract compared to normal sugars. Therefore the peak concentration of blood sugar is lower leading to a lower insulin response or low **glycemic index**. Sugar alcohols are also **noncarogenic** (i.e. unable to support cavity formation) since they are not broken down by microorganisms in the oral cavity. They have a **cooling effect** which is desirable in peppermints and in products such as chewing gums.

**Table 4.3**
**A list of some sugar alcohols**

| Sugar | Sugar alcohol | Relative sweeteners (%) (cf. sucrose) | Cooling effect |
|---|---|---|---|
| Glucose, fructose | Sorbitol | 55–60 | 4.4 |
| Lactose | Lactitol | 90 | 1.4 |
| Mannose | Manitol | 50–60 | – |
| Maltose | Maltitol | ~90 | 2.5 |
| Xylose | Xylitol | ~90 | 6.7 |

[c]Uronic acids are oxidation products formed by the selective oxidation of the C6–OH group. For example, oxidation of glucose yields glucuronic acid. Oxidation of gulose and mannose produces mannuronic acid and guluronic acids which are important building blocks for pectin and alginic acid (see later). Strong oxidizing agents can oxidize both ends (aldehyde and C6–OH) to form an aldaric acid. The sugar di-acids, such as glucaric acid, do not appear in nature.

The FDA allows foods which contain sugar alcohols to be labeled as "sugar free." Such foods can also be labeled as "Does not promote tooth decay," provided that the food does not contain other sugars. Sugar alcohols are used in confectionary products. They tend to produce a laxative effect if consumed in large quantities. Fortunately, the **laxation threshold** of sugar alcohols appears to differ for different individuals.

## 4.3  SUCROSE AND OTHER DISACCHARIDES

A disaccharide is formed when the (C2–C6) alcohol group of one sugar unit reacts with the anomeric carbon group of a second sugar. For example, maltose is formed when the C1 aldehyde group of glucose reacts with the C4 hydroxyl group of a second glucose molecule to produce a (1→4) glycosidic bond. Other **linkage types** (1→2, 1→3, 1→5, and 1→6) are possible between sugars. A glycoside bond is readily hydrolyzed by acid yielding the constituent monosaccharides. Familiar disaccharides include lactose (glucose + galactose) and sucrose (glucose + fructose). Maltose (Figure 4.4) (glucose + glucose) is produced during the digestion of starch.

### 4.3.1  Sources and Processing

Common sugar, table sugar, and saccharose are some of the names for sucrose.[1] Commercial sugar is extracted from sugar cane or sugar beet; the juice from such sources contains 11–17% sugar. This is readily concentrated to nearly 70% sugar by boiling to evaporate water. Sucrose then precipitates forming crystals and a brown syrup called molasses (in the US) or treacle (in the UK). White sugar is obtained by recrystallizing sucrose from solution. In the US, sugar is also produced from corn starch which is hydrolyzed using starch degrading enzymes (amylases). The product is initially glucose but is then isomerized to form a 50:50 mixture of **fructose** and **glucose** using glucose isomerase.

### 4.3.2  Physical Properties of Sucrose

Pure table sugar is >99% sucrose. When a hot and highly concentrated solution cools the sucrose precipitates. The molecules of sucrose form highly ordered arrays. Sucrose can also form a **glass** which is an amorphous solid. Glass formation is likely when a concentrated sugar solution is cooled rapidly and the solution "sets" before sucrose molecules have had time to form ordered arrays. Over time, a glassy material may transform slowly into a crystalline solid with the release of entrapped water.

### 4.3.3  Sucrose as a Food Ingredient

Crystalline sucrose is hygroscopic; that is, a material that absorbs moisture from the atmosphere. To prevent the uptake of moisture sucrose must be stored under conditions where

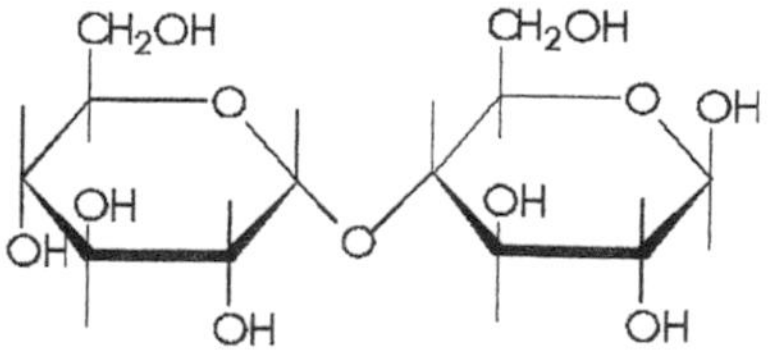

**Figure 4.4**  The structure of maltose. The two glucose units are linked by a $\alpha(1\rightarrow4)$ linkage. $\alpha$ shows that the C1–OH group involved in bonding points below the plane of the ring.

**Table 4.4**
**Grain sizes for sugars**

| Name | Approximate particle sizes (mm) |
|---|---|
| Coarse sugar | 1.0–2.5 |
| Medium sugar | 0.6–1.0 |
| Fine sugar | 0.1–0.6 |
| Icing sugar | < 0.1 |

the relative humidity is below 50%. Higher levels of moisture produce caking, during which, moist sugar particles become stuck together leading to a general increase in particle size and the formation of "lumps." Moist conditions also allow the growth of microbes. The relationship between relative humidity and amount of water absorbed by sucrose can be expressed via the **moisture adsorption isotherm**.

For convenience, sucrose is usually provided having specific grain sizes (Table 4.4). Apart from the difference in particle size, the different forms of common sugar are fundamentally the same.

Hydrolysis of sucrose leading to a mixture of fructose and glucose alters the sample's interaction with polarized light. A reaction called **inversion** occurs when sucrose is heated with acid. The production of invert sugar can also be catalyzed using the enzyme **invertase**, produced by yeast. Sucrose lacks an anormeric carbon group and is therefore a **nonreducing sugar**. By contrast, both glucose and fructose reduce $Cu^{2+}$ ion to $Cu^{+}$ which then precipitates as brown copper hydroxide. This reaction, promoted by high temperatures and high pH, is the basis for the Benedict or Fehling test for **reducing sugars**. Those sugars that lack a C1 aldehyde group and which do not react with copper sulfate solution are said to non-reducing sugars.

There are several reactions of sugars in food systems that are worthy of note. Earlier we referred to the *Maillard reaction* leading to brown pigments. Sugars will also form brown pigments in the absence of protein. When crystalline sugar is heated, it first melts, and then undergoes dehydration followed by polymerization to produce **caramel**. The details of **caramelization** are not fully understood. Caramel is an important ingredient and is widely used as a food color and flavor.

## 4.4  STARCH

### 4.4.1  Sources and Isolation

The main sources of commercial starch are commodity plants including cereals (corn, wheat) tubers (potato, cassava), and legumes (beans, lentils). Some plant seeds contain 80–90% starch by dry weight. The method for starch extraction is fairly similar regardless of the plant source. Key steps for preparing industrial starch involve soaking, grinding or wet milling, filtration to remove coarse plant materials, and drying to a fine powder. Commercial starches contain two polysaccharides, **amylose** and **amylopectin**, in a ratio of 0.25:0.75 (Table 4.5). High amylose corn (a natural mutant variety of maize) contains 75–80% amylose while glutinous rice and waxy maize contain over 90% amylopectin. The two components of starch can be separated by **selected precipitation** or **selective solubilization**. In the former case, starch is dissolved with an organic solvent, usually dimethylsulfoxide. Addition of butanol or iodine leads to the formation of an insoluble complex and precipitation of amylose. With selective solubilization, a slurry of starch is heated high temperature until the amylose

**Table 4.5**
**Starch granule sources, sizes, and morphology**

| Source | Shape/size | % Amylose | Swelling index (95°C) | Gelatinization temperature (°C) |
|---|---|---|---|---|
| Corn | | | | |
| Normal | P, 15 μm | 26 | 24 | 62–80 |
| Waxy | P, 15 μm | ~1 | 64 | 63–74 |
| HA | P, 15 μm | 80 | 6 | 65–87 |
| Wheat | P, 15 μm | 25 | 21 | 53–72 |
| Rice | P, 5 μm | 17 | – | 61–80 |
| Sorghum | P, 15 μm | 26 | 22 | 68–78 |
| Potato, cassava | S, 33–100 μm | 22 | 1000 | 56–69 |
| | S, 17 μm | | | |

The shapes of starch granules are polymorphic (P), spherical (S) or reniform (R).
Source: Pomeranz, Y. (1991). *Functional Properties of Food Components*. 2nd edn. Academic Press Inc., San Diego. pp. 24–79.

component dissolves. The mixture of soluble amylose and insoluble amylopectin can be separated by filtration.

### 4.4.2  STARCH AND GLYCOGEN—STORAGE POLYSACCHARIDES

Starch is the most abundant polysaccharide in the biosphere after cellulose. It serves as the energy reserve in higher plants. Bulk starch intended for use in food manufacturing comes from (1) cereals having 40–90% dry weight, including maize, wheat, rice, and sorghum; (2) tubers having 65–80% dry weight, including potatoes, cassava or manihots, arrowroot, and sweet potato; and (3) legumes with 30–70% dry weight, including peas, soya, and other legumes. Raw starch occurs as large particles called **starch granules** that are visible under a low power optical microscope. The sizes and shapes of starch granules vary in different species of plants (Table 4.5). Potato starch granules are relatively large and able to absorb water, readily swelling to twice their original size during cooking. Starch granules from cereals are generally small, compact, and difficult to swell during heating in water. Starches from the legumes exhibit properties that are intermediate compared to those of cereal and tuber starches (Table 4.5). **Animal starch** (glycogen) is found mainly in the liver at concentrations of 10% by mass or in the muscles at 1% (w/w). The breakdown of glycogen within muscle tissue has considerable impact upon meat quality (Chapter 12, Section 7).

### 4.4.3  AMYLOSE STRUCTURE

Amylose is a linear polymer of 1,4 linked $\alpha$-D-glucose units. Dry amylose adopts a **double helix** structure. When dissolved in hot water or cold dimethyl sulfoxide amylose adopts a random shape. Addition of iodine leads to an **inclusion complex** ($\gamma$-amylose) in which an array of iodine molecules is surrounded by an amylose helix. Approximately 20 mg of iodine is bound per 100 g of amylose. The blue–black coloration of the starch–iodine complex is due to the transfer of iodine from a polar water medium into the nonpolar interior of an amylose helix. The absorbance maximum for iodine changes from 450 nm (yellow, brown) in a polar medium to 620 nm (blue, violet) in a hydrophobic environment. Amylose also forms inclusion complexes with other **nonpolar solutes**, for example, butanol, fatty acids, and flavor compounds. This is one explanation for how starchy food products interact with a range of nonpolar molecules. Enzymatic hydrolysis of amylose leads to shorter chain products with low iodine binding. The reduction in iodine binding capacity is an effective way of monitoring amylase activity.

### 4.4.4 Amylopectin Structure

Amylopectin consists of linear 1,4 linked $\alpha$-D-glucose units with $\alpha(1\rightarrow6)$ branch points every 15–20 glucose units. The **cluster model** for amylopectin shows this to be arranged like a fern. The branching has a profound effect on the properties of amylopectin. **Chain branching** prevents the extensive helix formation needed for iodine binding. Amylopectin binds about 2 mg of iodine compared to 20 mg iodine per 100 g amylose. The percentage of amylose in 100 g of starch is generally estimated from the amount of iodine bound ($Y$ mg) using the approximate formula, % amylose $= (Y \times 100/20)$. Chain branching also retards gel formation and therefore amylopectin is virtually nongelling. Finally, a high degree of branching improves the solubility of starch. Glycogen is more highly branched than amylopectin.

### 4.4.5 Interactions Between Starch Granules and Water

The starch granule is a macromolecular complex of amylose and amylopectin. The microstructure of a starch granule shows a series of **concentric rings** believed to represent **crystalline** and **amorphous** regions. These appear to be the result of regularities in amylopectin chain branching. Starch granules can be classed a **type A** (cereal starch) or **type B** (potato starch) which have high and low levels of internal order, respectively. **Type C** (cassava) starch has intermediate order. The different types of starch can be differentiated based on their X-ray diffraction patterns. Type A–C starches also differ in terms of their cold water solubility, heats of hydration, and swelling index.[2]

The interaction of powdered starch with water vapor is described by a classic **Langmuir absorption** isotherm. A plot of relative humidity versus percent water bound shows **hysteresis**. The water content of a starch sample depends on sample history. Type A starch (cereal starches) absorbs less water vapor compared to type B.

Native starch is insoluble in cold water. Adding starch to cold water produces a dispersion. The suspended starch granules are insoluble and can be recovered by simple physical processes such as filtration. The size, shape, and general integrity of the starch granules remain unchanged in cold water. This description is also true for raw products such as potatoes or wheat dough.

When starch is heated gradually at 1.5°C/min there follows a series of physical changes at a definite temperature which is unique for each kind of starch. At the **gelatinization temperature** there is a sudden increase in water absorption leading to 100% increase in the volume of starch granules. The starch suspension turns from a cloudy opaque hue to a partially transparent solution. Amylose is released from starch granules. Amylopectin, due to its branched profile, is retained within the partially disrupted granules, sometimes called ghosts. There is also an increase in solution viscosity. Finally, there is a loss of birefringence, that is, in the characteristic Maltese cross patterns formed when native starch is illuminated with plane polarized light. These physical changes are a sign of the process of **gelatinization** and they occur at the gelatinization temperature. Differential scanning calorimetry and viscometry are convenient methods for assessing the gelatinization temperature. Viscosity or the birenfrigent properties can be measured whilst a starch suspension is heated at a constant rate. A suspension of swollen starch granules within a matrix of dissolved amylose polymers is called **a paste**. This term is applied to a variety of systems with large variations in water content. The behavior of starches during cooking, under the influence of heating and shear, is called the **pasting characteristics**. Gelatinized starch serves as a thickening agent for sauces and soups. Pasting behavior is also important for processing cereal based products (baking bread and cookies).

Cooling a hot starch solution leads to **gelation**. Chains of amylose molecules become hydrogen bonded along their lengths forming **junction zones**. The resulting polymer–polymer interactions become disrupted whenever interacting amylose chains encounter the partially intact granule giving rise to an **interruption zone.** Repeated breaking, and reformation of junction and interruption zones leads to a **3-dimensional gel network** capable of entrapping large amounts of water. The formation of gel networks is therefore said to depend on the balance of polymer–polymer and polymer–solvent interactions. Food gels provide **water holding** and **texture.** Both interruption and junction are needed for **gelation.** In the absence of interruption zones polymer–polymer contacts lead to rope-like structures or fibrils.

The process of gelation is also called **retrogradation** to indicate partial return of a granule-like, ordered structure within the starch gel.[3] Retrogradation provides a possible explanation for the hardening of freshly cooked starchy foods with time. Baked products, including bread and cookies, have a soft texture shortly after cooking. However, freshly baked products increase in hardness within a few minutes of cooling, due to the gelation of amylose. The much slower and prolonged **staling** which occurs over a period of days is ascribed to amylopectin retrogradation within the starch granule ghosts. Additives such as butter and fatty acids can modify the gelation characteristics of starch thereby producing an anti-staling effect.

## 4.5   PECTIN

Pectin is an exo-polysaccharide from terrestrial plants. It functions as an **intercellular cement** and helps to hold plant cells together. The texture of nonwoody vegetative plant material such as fruits and tubers is dependent on the pectin content. Commercial pectin is isolated from citrus fruit (apples, lemons, limes) or apple rind obtained from fruit juice processors. The manufacture of pectin and other acidic **industrial polysaccharides** involves alkali extraction followed by precipitation using calcium or ethanol (see detailed description for alginate, Section 4.6). Most terrestrial plants have pectin in their cell walls.

The structure of pectin is based on two types of monosaccharides. First, there are repeated block sequences of poly-**galacturonic acid methyl ester** (GGGGG). Block sequences of the neutral sugar **rhamnose** (RRRRRR) also occur along with small amounts of mixed sequences.[4] The G-units of newly synthesized pectin are in a fully esterified state. Pectin extracted from unripened fruit has a high degree of methylation and is called **high methoxy pectin** (HMP). During fruit maturation and ripening, HMP becomes demethylated leading to **low methoxy pectin** (LMP). From a technical viewpoint, the range of pectic substances includes protopectin (100% esterifed), pectinic acid (partially esterified), and pectic acid (fully de-esterified). The two main commercial pectin grades, HMP or LMP, have different gelation characteristics.

The conformational characteristics of poly-G and poly-R sequences are very different. The $\alpha$-D-galacturonic acid residues are joined by $\alpha$-1,4 axial linkages. This **linkage conformation** leads to sugar units being stacked one above the other. The space between adjacent residues forms a **ligand binding pocket** (Figure 4.5). By contrast, $\beta$-1,4 linked rhamnosyl units possess a backbone conformation similar to that of cellulose. As described elsewhere the extended conformation associated with $\beta$-1,4 linked glucans accounts for their overall rigid structure, large end–end distance, and ability to produce highly viscous solutions (Chapter 7, Table 7.7).

### 4.5.1   GELATION OF PECTIN

Pectin powder can be dispersed with water if care is taken to avoid clumping. When heated in solution the helical pectin molecule denatures, loses its native tertiary structure, and thereafter adopts a random shape. Some prevailing models for the random conformation are discussed in Chapter 7, Section 7.6. For example, solution viscosity increases in proportion to the polymer molecular weight and characteristic ratio. As the temperature of a polysaccharide solution is reduced, pectin forms a gel as a result of forming **junction zones** and **interruption zones.**

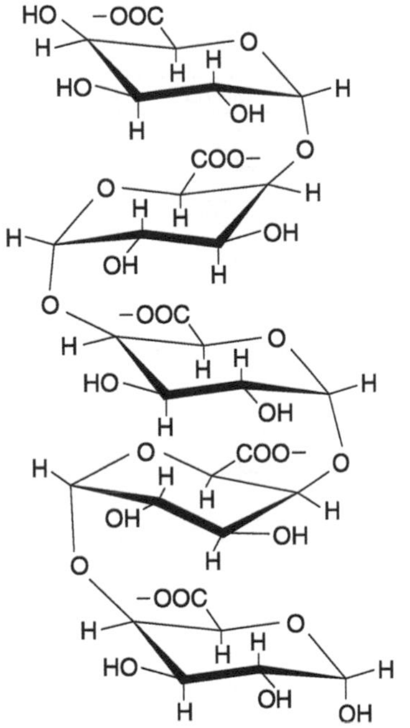

**Figure 4.5** Conformational structure of poly-galacturonic acid residues of pectin.

In the case of HMP, junction zone formation requires sucrose (~60 g/100 ml of solution) and a low pH. The high concentration of sucrose is thought to strengthen **hydrophobic interactions** between methylated pectin chains. Decreasing pH increases polymer chain–chain interactions as anionic galacturonate residues ($-COO^-$) are transformed in the non-ionized (COOH) state. Whilst adjacent carboxylate groups will repulse each other, the non-ionized carboxylic acid groups can be expected to attract each other via H-bond formation. LMP forms gels in the presence of calcium ions. Since there is no requirement for sugar, LMP are suitable as a basis for low calorie products intended for diabetics.

Methods for assessing pectin gel strength include (1) the jelly grade or SAG method, (2) texture analysis using the Voland Stevens Analyzer, and (3) determination of setting temperature.[5]

### 4.5.2 Pectin and Texture in Plant Foods

Pectin is degraded by enzymes naturally present in plant cells leading to softening or a loss of texture during fruit ripening. The reaction requires two enzymes: **pectin methyl esterase** (PME) and **polygalacturonase** (PGase). The first enzyme catalyses the de-esterification of pectin transforming HMP to LMP. This is followed by the action of PGase which catalyses the hydrolysis of the pectin chain. By working together, PME and PGase cause the de-polymerization of pectin. The role of pectin in fruit texture is discussed further in Chapter 11.

## 4.6 ALGINATE

**Alginic acid** is a structural polysaccharide isolated from brown seaweed (genus *Phaephacea*) that grows wild in the temperate zone (average ocean temperature 4–18°C) at sea depths

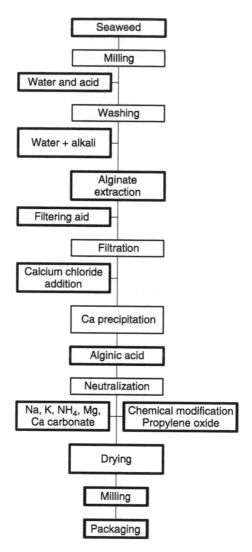

**Figure 4.6**   Processing and production of alginate.

of < 50 m. Much of the seaweed for alginate extraction (*Laminaria* sp. and *Macrocystis purifer*) comes from the coasts of the USA, South America, South Africa, and Japan. There is no concerted **agronomic production** of seaweed, although harvesting from natural stocks leads to concerns about sustainability. Systems for aqua culture are therefore being considered. **Processing** dried seaweed feedstock for alginate extraction is summarized in Figure 4.6. After milling, seaweed powder is washed with acid, in which alginic acid is insoluble. The cleaned material is extracted with sodium hydroxide and precipitated by calcium. Depending on the commercial requirements, calcium alginate can be resolubilized with sodium, potassium or ammonium hydroxide to form sodium, potassium or ammonium alginate for commercial use.[6]

### 4.6.1   STRUCTURE OF ALGINATE

Alginate is a block homopolymer of $\beta$-D-*mannuronic acid* (poly-M), $\alpha$-L-*guluronic acid* (poly-G), and mixed sequences of M and G (Table 4.6). The poly-M sequences adopt the extended helix secondary structure associated with 1,4-$\beta$-D-glucan units of cellulose

---

**Table 4.6**
**Mannuronic acid and guluronic acid residues in alginate\***

| Type of seaweed | % poly-M | % mixed (MG) | % poly-G |
|---|---|---|---|
| L. hyperborea (stem) | 17 | 26 | 57 |
| L. hyperborea (leaf) | 36 | 38 | 26 |
| L. digitata | 43 | 32 | 25 |
| L. japonica | 48 | 36 | 16 |
| M. pyrifera | 38 | 46 | 16 |

*L = Laminaria. M = Macrocystis.

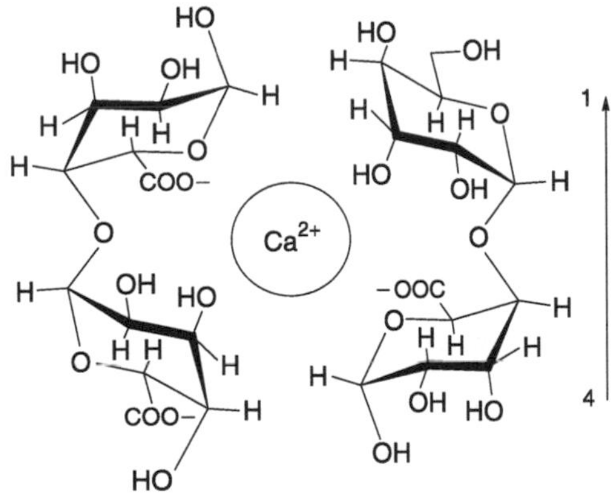

---

**Figure 4.7** A model for the egg-box structure formed when alginate binds $Ca^{2+}$ ions via polyguluronic acid residues.

(Chapter 7). However, the poly-M sequences exhibit greater solubility compared with cellulose which consists only of glucose subunits. Mannuronic acid residues differ from glucose in two principal ways: C2 for mannose has an axial OH group rather than the equatorial OH of glucose and C5 has a charged $COO^-$ as compared with the $CH_2$–OH residue.

The guluronic acid block sequences of alginate are liked by the same $\alpha$-1,4 (axial–axial) linkages described for pectin (Figure 4.5). As described before, in the absence of added minerals the ligand binding pockets are occupied by water.

## 4.6.2 GELATION OF ALGINATE

Alginate molecules adopt an essentially random conformation in solution. The resulting viscosity increases as a function of concentration, molecular weight, characteristic ratio, and polymer–polymer interactions in the semi-dilute regime (see Chapter 8). Addition of alkali metal ions ($Mg^{2+}$, $Ca^{2+}$, $Sr^{2+}$ or $Ba^{2+}$) induces alginate gelation. **Junction zones** form due to interchain cross-linking by metal ions. For example, $Ca^{2+}$ cross-links four guluronic acid residues from adjacent poly-G chains, leading to the so-called **egg-box structure** (Figure 4.7). The ligand binding pockets are virtually the same as described for pectin.

The **interruption zones** for alginate gels occur at the poly-M sequences. Since these do not bind metal ions, changing the level of M sequences alters the gelation of alginate. Material with a high G:M ratio forms hard and brittle gels, whereas alginate with a low G:M ratio produces soft and elastic gels (Table 4.6). The G:M ratio varies with the age of seaweed, botanical species, portion of plant used, distance from the sea shore, and depth of water. Mature seaweed fronds or plants harvested nearer in-shore possess a higher G:M ratio as a result of their higher maturity and/or greater adaptation to a strong surf. Newly synthesized alginates consist almost entirely of poly-M sequences which are then subsequently converted to guluronic acids by the enzyme **mannuronic acid-5C-epimerase**.

### 4.6.3 ALGINATE APPLICATIONS

Alginates are truly versatile ingredients. Their **gelation** properties are exploited for products such as **restructured** or re-formed vegetables, meat or fish. An aqueous extract of meat, fruit or onion flavor is added to alginate and the mixture is gelled by adding $Ca^{2+}$. The resulting restructured meat, onion or fruit can be colored, sliced, and cooked as necessary. Decorative cherries and fried onion rings are some popular alginate products. Alginates can also be used in puddings and deserts, and in cold prepared bakery fillings. The **thickening** action is applied in tomato ketchup and sauces, soups, and milk shakes. Alginate has a **stabilizing** function in mayonnaise, whipped topping, and ice cream probably as a result of its thickening action. The **film forming** properties of alginate come into play when it is used as a glaze or fish coating.

## 4.7  XANTHAN AND MICROBIAL EXO-POLYSACCHARIDES

Xanthan was the first microbial polysaccharide approved for use in foods in 1969–1970 by the Food and Drug Administration (USA). This was accepted as a food additive (E415) by the European Union 1974. Xanthan is produced by submerged culture of the microbe *Xanthomonnas campesteris* grown on minimal media. Processing involves (1) fermentation, (2) centrifugation and pasteurization of the growing broth, (3) isopropanol precipitation of xanthan, and (4) spray or drum drying, followed by (5) milling and packaging.

The structure of xanthan consists of a $\beta$-1,4 linked D-glucose central chain supporting a trisaccharide chain (3,1-$\alpha$-D-mannopyranose-2,1-$\beta$-D-glucouronic acid-4,1-$\beta$-D-mannopyranose) at alternating glucose units. The main chain shows a high characteristic ratio and rigidity associated with cellulose-like polymers. The presence of side-hairs prevents strong polymer–polymer interaction leading to enhanced solubility compared with cellulose. The presence of hairs also prevents junction zone formation and consequently xanthan is non-gelling. In solution, xanthan forms a rigid rather than a random structure associated with other dissolved polysaccharides.[7] The structure of xanthan is shown in Figure 4.8.

### 4.7.1 THICKENING BY XANTHAN

Xanthan is the quintessential thickening agent. Its intrinsic viscosity (7500–12,500 ml/g) is considerably higher than values recorded for alginate ($\sim$800 ml/g) or starch (80 ml/g). Low concentrations (0.02–0.6 % w/v) of xanthan are able to generate high viscosities. Xanthan solutions show high yield value. This means that relatively high stress is needed to produce a shear rate of $0.01\,s^{-1}$. Xanthan solutions also exhibit psuedoplasticity (shear thinning).

Some characteristics of xanthan which are important for industrial applications include its **high stability** with respect to pH, heat, and reaction to enzymes. Xanthan is suitable as a thickening agent over a pH range of 1.5–13. There is no sign of thermal degration of xanthan at temperatures of 0–80°C. However, at 115°C xanthan is degraded with a half-life of about 200 min. The intestinal digestibility of xanthan is low, therefore this polysaccharide is suitable for low calorie products. Owing to its low digestibility, xanthan functions as dietary fiber. Like all other negatively charged polysaccharides, xanthan will interact with proteins at low pH. Xanthan forms gels in the presence of galactomannans.[8]

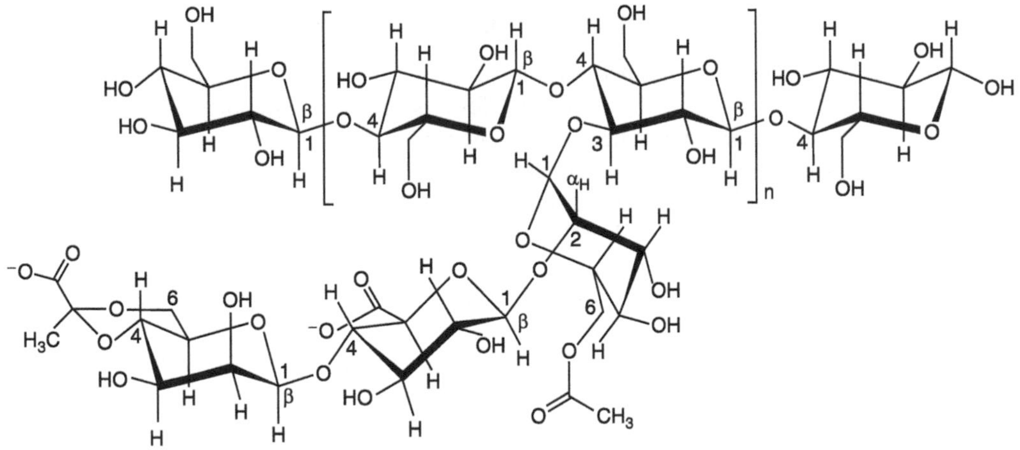

**Figure 4.8** Structure of xanthan. Bracket shows backbone chain with trisaccharide "hair."

### 4.7.2 APPLICATIONS OF XANTHAN

Xanthan (**E514**) is a permitted thickening agent in many foods. It functions as a **stabilizer** and imparts smooth texture and creamy consistency. Xanthan (0.2–0.6% w/v) is added to pourable dressings (mayo, salad dressing) and dairy products (0.02–0.06% w/v) such as milk shakes, ice cream, and whipped desserts. Xanthan is also found in sauces, gravies, and dry mixes.

### 4.7.3 OTHER MICROBIAL EXO-POLYSACCHARIDES

Other microbial polysaccharides have been developed in addition to xanthan. **Gellan gum** (E418) is a linear polysaccharide produced by microbial fermentation using *Pseudomonas elodea*. The structural unit for gellan gum is a linear chain ([Glc. Glc.A. Glc. Rha]$_n$) and consequently this is a gelling agent at concentrations of 0.1% (w/v). Acylated gellan can produce hard brittle gels. Very low concentrations produce "fluid gels" or material with a consistency of spread. Commercial gellan gum (Kelcogel) is produced by Kelco Ltd. **Wellan gum**, a non-gelling polysaccharide produced by *Alcligenes* sp. is structurally similar to gellan but possesses a rhamnose "hair" on the third residue [Glc. Glc.A. Glc. (Rha). Rha]$_n$. Finally, the other microbial polysaccharide worth mentioning is **rhamsan gum** which, being branched, is also non-gelling.[9]

## 4.8 GUAR, TARA, AND LOCUST BEAN GUM

The seed gums or **galactomannans** are the most abundant group of storage polysaccharides after starch. They are mostly extracted from leguminous seeds. The three most important commercial seed gums are **guar gum**, **tara gum**, and **locust bean gum**. The chemical architecture of the seed gum is based on a $\beta$-1,4 linked mannose backbone with periodic occurrences of galactose side-chains or "hairs." With decreasing content of galactose residues the solubility of the seed gums decreases. For instance, the mannose:galactose ratio is 2:1 (guar), 3:1 (tara), and 4:1 (locust bean gum). Guar gum and tara gum exhibit high and low solubilities in cold water. By contrast, locust bean gum dissolves only after heating. The plant seed gums are essentially non-gelling. Their rheological behavior can be understood in terms of structure–functionality relations discussed for xanthan and further explained in Chapter 7.

## 4.9  FUNCTIONALITY OF POLYSACCHARIDES

Polysaccharide functionality is defined as any physical or chemical property, in addition to their nutritional role, which is of interest to food scientists. The principal role of polysaccharides is as **gelling** and **thickening** agents. The ability to stabilize dispersions is the result of **gelation** or **viscosity** modifying action. Gelation involves the formation of 3-dimensional gel networks (starch, pectin, alginate) via the formation of polymer–polymer (junction zones) and polymer–solvent (interution zones) bonding. Branched polysaccharides having side chains (xanthan and seed gums) are non-gelling since the presence of "hairy" regions interfere with junction zone formation. The extent of thickening and other rheological characteristics is related to polymer end–end distance as described in the section on food rheology.[10,11] The solution behavior of polysaccharides is further described in Chapters 7 and 8.

## References

1. Beckett, S., *The Science of Chocolate*, Royal Society of Chemistry, Cambridge, UK, 17–23, 2000.
2. Whistler, R. L., *Carbohydrate Chemistry for Food Scientists*, Egan Press, St Paul, MN, 1997.
3. Miles, M. J., Morris, V. J., Orford, P. D., and Ring, S. G., The roles of amylose and amylopectin in the gelation and retrogradation of starch, *Carbohydrate Res.* 135, 271–281, 1985.
4. De Vries, J., Repeating units in the structure of pectin, in *Gums and Stabilizers for the Food Industry*, Vol. 4, Philips, G. O., Wedlock, D. J., and William, P. A., Eds, IRL Press, Oxford, 25–29, 1988.
5. Ikkala P., Characterization of pectins: A different way of looking at pectins, in *Gums and Stabilizers for the Food Industry*, Vol. 3, Philips, G. O., Wedlock, D. J. and William, P. A., Eds, IRL Press, Oxford, 253–267, 1986.
6. Iverson, A., *Thickening and Gelling Agents for Food*, Blackie Academic and Professional, London, 1992.
7. Milas, M., Rinaudo, M., and Tinland, B., Role of the structure on the rheological behavior of xanthan gum, *Gums & Stabilizers* 3, 637–644, 1986.
8. Cuvelier, G. and Launey, B., Viscoelastic properties of Xanthan–carob mixed gels, *Gums & Stabilizers* 3; 147–158, 1986.
9. Pettitt, D. J., Recent developments—future trends, *Gums & Stabilizers* 3, 451–463, 1986.
10. Mitchell, J. R. and Ledward, D. A., *Functional Properties of Food Macromolecules*, Elsevier Applied Science, New York, 1986.
11. Stephen, A. M., *Food Polysaccharides and Their Application*, Marcel Dekker Inc., New York, 1995.

# 5

# Lipids and Fat Replacers

## 5.1  INTRODUCTION

Lipids are described in this chapter with emphasis on their role as food **ingredients**. The sources of commodity oils are reviewed in Section 5.1.2. Processing methods are discussed in Section 5.2. The organic chemistry of fats and oils is described in Sections 5.3 and 5.4. The functional group chemistry outlined in Section 5.4 is key to understanding many aspects of fats and oils technology. For example, **oxidation** is the main route for fat and oil deterioration. **Iodine binding** provides a simple method for assessing the degree of unsaturation. **Hydrogenation** is the basis for manufacturing hardened vegetable oils for use as shortening and margarine. Sections 5.5 to 5.7 consider the physical chemistry of fats and the relationship between **crystallization**, melting, and mechanical properties of confectionary foods and desserts, including chocolate and ice cream.

### 5.1.1  DEFINITION AND CLASSES OF LIPIDS

A lipid is any material from a biological source that dissolves with organic solvents including hexane, acetone, petroleum ether or paint thinner. The four major classes of lipids are listed in Table 5.1. The **triacyglycerols** (TAGs), formerly known as triglycerides, are most well known for their dietary role as the most calorie-dense food material. **Cholesterol** is now well established in the collective psyche as a contributary factor for cardiac vascular disease. The term oils and fat describe TAGs in the liquid and solid state, respectively. Everyday experience shows that oils and fats are interconvertible via melting as the surrounding temperature is raised or lowered. An oil in the summer months may become a fat in winter; an oil in the tropics may be a fat in temperate climates. Nevertheless, owing to their generally high degree of saturation, animal derived TAGs tend to be fats whilst plant and fish derived TAGs are generally oils.

### 5.1.2  SOURCES OF EDIBLE OILS

Some of the major sources for edible commodity oils are listed in Table 5.2. The data show that **soybean oil** from *Glycine max* is the most important vegetable oil followed by **palm oil** from *Elaeis guineensis*. Also readily available from supermarkets are oils from sunflower (*Helianthus annuus*), **rapeseed** or canola (*Brassica napus* or *B. campestris*) and **corn** (*Zea mays*). The major plant sources of oils are legumes (soyabean, peanut, sesame seed), cereals (corn germ), oilseeds (rapeseed, cottonseed), and tree crops (palm, coconut, cacao). Animal fats are also important though not listed.

    **Olive oil** is also exceptional for being obtained from tree seeds. Shear nut oil is also important in tropical areas. Vegetable oils are used to cooking, frying, and as salad dressing. Animal **fats** include *beef tallow* or rendering fat, *butter fat*, as well as *fish oil*. Consistent with our emphasis on general principles we will not focus on specific commodities.

**Table 5.1**
**Characteristics of lipids**

| Lipid class | Comments |
|---|---|
| Waxes | Esters of long chain alcohols and fatty acids, found as coatings in plants |
| Sterols | Cholesterol is an example of a steroid. It as needed for proper cell membrane function |
| Phospholipids | Forms cell membranes. Phospholipids form "gums" in crude oil. Used as emulsifiers in food manufacturing |
| Triacylglycerol | Edible oil or fat |
| Vitamins | Vitamins A, E, D, and K are sometimes classed as lipids |

**Table 5.2**
**World consumption of vegetable and marine oils (USDA, 1999)**

| Oil sources | Million metric tons |
|---|---|
| Soybean oil | 24.5 |
| Palm oil | 21.2 |
| Rapeseed (canola) oil | 13.3 |
| Sunflower seed oil | 9.5 |
| Peanut oil | 4.3 |
| Cottonseed oil | 3.7 |
| Coconut oil | 3.2 |
| Olive oil | 2.4 |
| Marine (fish) oil | 1.2 |
| **Total** | 85.7 |

However, soybean oil offers an excellent case study for discussion of the edible-oil industry. Figure 5.1 shows the scale of soybean oil production and uses in the US (1999). Clearly, soybean oil is the most important oil and most of this product is used for food applications. It is noteworthy that 260 thousand metric tons of soybean oils were used for industrial applications, including resins and plastics (52 TMt), ink (49 TMt), paint and varnishes (18 TMt), fatty acids, soap, and feed (164 TMt), biodiesel (4 TMt), and as solvent (0.5 TMt).

## 5.2   PROCESSING EDIBLE OILS

Some of the major steps for plant seed oil processing involve the handling of raw materials, oil extraction, degumming, refining, decolorization using bleaching clay, deodorization or removal of volatile off-flavors using steam injection, and the optional steps of **hydrogenation** and **trans-esterification**. The oil extraction process alone contains six separate operations, as follows.[1]

(1) **Selection and cleaning** of soybeans. For quality assurance, the processor monitors seed maturity, sprouting, degree of heat damage, and the oil and protein content of seeds which are generally inversely related.

(2) **Drying and storage**. The moisture content of seeds is reduced to prevent spoilage. Storage temperature, degree of movement, and general transfer of old and new feedstock throughout the storage area are some important considerations.

(3) **Cracking and dehulling**. Soybean are broken into six to eight parts after which the hull is removed.

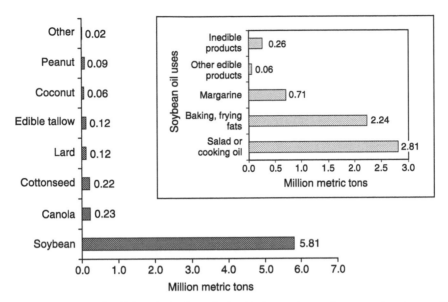

**Figure 5.1** Consumption of edible oils in the US (1999). Insert shows the uses of soybean oil. Drawn using data from United Soybean board.

(4) **Conditioning** and **flaking**. Moisture is controlled to yield pliable beans for flaking.
(5) Solvent extraction. Soybean flakes (18% oil) are brought into contact with hexane at 120–140°F in countercurrent flow. The residual edible oil is usually reduced to 0.3–0.4%.
(6) Desolventization of seed flakes. Soyflakes (30% hexane) are subjected to sparging with steam to vaporize the hexane. The moisture content of seed flakes is reduced to 20% w/w. Some desolventization machines perform toasting and drying functions at the same time.

The end product of soybean extraction is crude soybean oil and soybean meal. The former is then subjected to refining operations as shown in Figures 5.2 and 5.3. The soybean meal is, of course, a vital feedstock for producing soybean proteins.

## 5.3 FATTY ACIDS

The TAGs contain glycerol and three acyl groups. The three hydroxyl groups of glycerol are each esterified with a carboxylic acid. We will recall from organic chemistry that carboxylic acids (e.g., acetic acid) react with alcohols to form an ester with the release of one molecule of water. The reaction shown in Eq. 5.1 is reversible with an equilibrium constant very close to 1:

$$Ac.COOH + ROH \rightleftharpoons Ac.CO.OR' + H_2O. \tag{5.1}$$

Esterification has an impact on the stability of fats and oils. **Phospholipids** resemble TAGs except that the third hydroxyl group of glycerol is attached to a phosphate moiety which is then attached to a further group.

The three carboxylic acids forming part of a triacylglyceride have the general formula **$CH_3(CH_2)_{n'}COOH$** where $n'$ (the number of $CH_2$ groups) ranges from 10 to 20 and the total

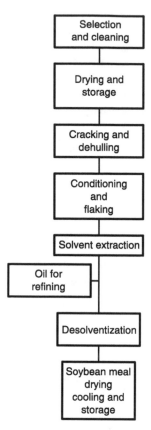

**Figure 5.2**   General processing steps for soybean oil.

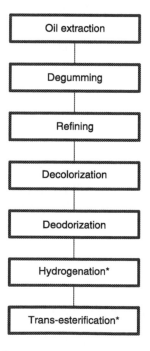

**Figure 5.3**   The refining of soybean oil.

number of carbon atoms ($n$) is 12 to 22. Long carbon chains have decreased solubility in water and a tendency to dissolve in oil, leading to the nickname, fatty acid.

### 5.3.1  NOMENCLATURE OF FATTY ACIDS

The fatty acids have three kinds of names. The **trivial** or common names for fatty acids having 2, 4 or 6 carbon atoms are acetic acid, butyric acid or caproic acid. The corresponding **systematic** names, also known as the IUPAC (International Union of Pure and Applied Chemistry) names, are ethanoic acid, butanoic acid, and hexanoic acid. Finally, there are the **abbreviated** names C2:0, C4:0, and C6:0. The first digit indicates that these fatty acids contain 2, 4, and 6 carbon atoms, respectively. The zero digit shows that each of these fatty acids has no double bond. The names for some common fatty acids are summarized in Table 5.3.[2]

### 5.3.2  SATURATED AND UNSATURATED FATTY ACIDS

**Mono-unsaturated fatty acids** contain one double bond. The position of the double bond within the carbon chain is indicated using the delta system. For illustration, **palmitoleic acid** has the abbreviated formula C16:$1^{(\Delta 9)}$. The formula for oleic acid is C18:$1^{(\Delta 9)}$. The delta-9 notation shows that both fatty acids possess one double bond located between C9 and C10. **Polyunsaturated fatty acids** possess two or more double bonds within their structure. Saturated fatty acids (S) contain no double bonds.

Unsaturated fatty acids are also classed according to the **omega system**. The omega number is the location of the first double bond counted *from* the $CH_3$ end of the fatty acid chain. Table 5.3 shows that most polyunsaturated lipids contain either **omega-3** or **omega-6** fatty acids (linoleic and arachidonic acid). Diets high in omega-3 or omega-6 fatty acids may help to prevent cardiovascular disease. It is not certain how these fatty acids achieve the health role. They may have other beneficial health effects through their conversion to **prostaglandins** and **leukotrienes**, which have anti-inflammatory and anti-hypertensive action in the body.

Alternative arrangements around a C=C double bond lead to **geometric isomers** for unsaturated fatty acids (Figure 5.4). *Cis* **fatty acids** are the most common form of fatty acids in nature. They can rearrange to form *trans* **fatty acids** during the hydrogenation of edible oils. Oleic acid (18:1 *cis* $\Delta 9$) gives rise to the *trans* fatty acid isomer **elaidic acid** (18:1 *trans* $\Delta 9$) during the partial hydrogenation of vegetable oils (see below). Small quantities of *trans* fatty acid occur naturally in ruminants (cows, sheep, buffalo) due to **biohydrogenation** as a result of anaerobic fermentation in the rumen. This process may account for the small level (2%) of *trans* fatty acids found naturally in butter. By comparison, hydrogenated vegetable shortening contains 30–40% *trans* fatty acids.[3] Figure 5.4 shows that a *trans* fatty acid has two hydrogen atoms on *opposite* sides of the double bond but the *cis* fatty acid has two hydrogens on the same side. This small change has significant consequences for the overall shape of a fatty acid, which, in turn, affects its physical properties and how it behaves in the circulation. *Cis* fatty acids are sometimes called **Z fatty acids** whilst *trans* fatty acids are designated as **E fatty acids**. The Z and E comes from the German terms *Zusammen* and *Entegegen* meaning same and opposite side, respectively.

## 5.4  CHEMICAL REACTIVITY OF FATS AND OILS

### 5.4.1  OXIDATION

Fat oxidation occurs when activated oxygen species react with unsaturated fatty acids leading to off-flavors (Chapter 9). The process is a free radical chain reaction. The key

**Table 5.3**
**The fatty acids and their trivial and systematic names**

| Fatty acid | Trivial name (symbol) | Systematic name | Mpt(°C) |
|---|---|---|---|
| **Saturated** | | | |
| C4:0 | Butyric acid (B) | Butanoic acid | −7.9 |
| C:6:0 | Capronic acid (C) | Hexanoic acid | −3.4 |
| C8:0 | Caprylic acid (Oc) | Octanoic acid | 16.3 |
| C10:0 | Capric acid (D) | Decanoic acid | 31.2 |
| C12:0 | Lauric acid (La) | Dodecanoic acid | 43.9 |
| C14:0 | Myristic acid (M) | Tetradecanoic acid | 54.1 |
| C15:0 | | Pentadecanoic acid | 62.7 |
| C16:0 | Palmitic acid (P) | Hexadecanoic acid | 69.9 |
| C17:0 | | Heptadecanoic acid | |
| C18:0 | Stearic acid (S) | Octadecanoic acid | |
| C20:0 | Arachidic acid (Ad) | Eicosanoic acid | 75.5 |
| C22:0 | Behenic acid (Bn) | Docosenoic acid | 79.9 |
| C24:0 | Lignoceric acid | Tetracosanoic acid | |
| **Monounsaturated** | | | |
| C14:1 (*cis* Δ9) | Myristoleic acid | *cis*-9-Tetradecenoic acid | |
| C16:1 (*cis* Δ9) | Palmitoleic acid (Po) | *cis*-9-Hexadecenoic acid | 0.5 |
| C18:1 (*cis* Δ9) | Oleic acid (O) | *cis*-9-Octadecenoic acid | 14.0 |
| C18:1 (*trans* Δ9) | Elaidic acid | *trans*-9-Octadecenoic acid | 43.7 |
| C18:1 (*cis* Δ11) | Vaccenic acid | *cis*-11-Octadecenoic acid | |
| C20:1 (*cis* Δ9) | Gadoleic acid | *cis*-9-Eicosenoic acid | |
| C22:1 (*cis* 13) | Erucic acid | *cis*-13-Docosenoic acid | |
| C22:1 (*cis* 11) | Cetoleic acid | *cis*-11-Docosenoic acid | |
| C24:1 (*cis* 15) | Nervonic acid | *cis*-15-Tetracosenoic acid | |
| **Polyunsaturated*** | | | |
| C18:2 n-6 | Linoleic acid (L) | *cis*-9,12-Octadecadienoic acid | −5.5 |
| C18:3 n-3 | Linolenic acid (Ln) | *cis*-9,12,15-Octadecatrienoic acid | −10.5 |
| C18:3 n-6 | γ-Linolenic acid | 6,9,12-Octadecatrienoic acid | |
| C18:4 n-3 | Stearidonic acid | *cis*-6,9,12,15 Octadecatetraenoic acid | |
| C20:4 n-6 | Arachidonic acid (An) | *cis*-5,8,11,14-Eicosatetraenoic acid | −49.5 |
| C20:5 n-3 | Timnodonic acid (EPA) | *cis*-5,8,11,14,17-Eicosapentaenoic acid | |
| C22:5 n-3 | | *cis*-7,10,13,16,19-Docosapentaenoic acid | |
| C22:5 n-6 | DPA | *cis*-4,7,10,13,16-Docosapentaenoic acid | |
| C22:6 n-3 | DHA | *cis*-4,7,10,13,16,19-Docosahexaenoic acid | |

*Alternative abreviated notation for polyunsaturated fatty acids, e.g., C18:2 n-6 refers to an 18 carbon fatty acid having two double bonds the last of which occurs at C6 counting from the methyl end of the chain; cf. description of omega 6 fatty acids. Source: Danish Veterinary Food Administration http://www.foodcomp.dk/ fcdb_aboutfooddata_proximates.htm#Protein

intermediates are **hydroperoxides**, which degrade to volatile **aldehydes** and **ketones** with strong off-flavors. The most popular methods for assessing oil oxidation include the **thiobarbituric acid** (TBA) test and the measurement of **peroxide value** (PV).

The TBA test is used widely in studies of fat oxidation in a range of foodstuffs. TBA reacts primarily with molanyl aldehyde produced from fatty acids having three or more unsaturated double bonds. TBA also reacts with a wider range of products, not all of which are lipid oxidation products: ketones, ketosteroids, acids, esters, sugars, imides and amides, oxidized proteins, pyridines, pyrimidines, and vitamins.[4,5] PV is equal to milliequivalents of iodine released when lipid hydroperoxides react with hydrogen iodide (HI). The concentration of free

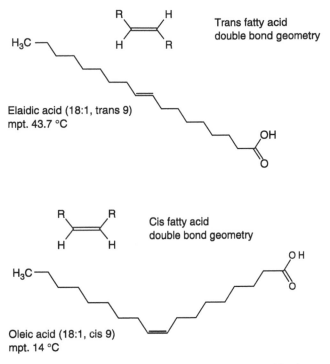

**Figure 5.4** Fatty acid geometric isomerism. Showing the arrangement of hydrogen atoms at the double bond for *cis* fatty acids and *trans* fatty acids.

iodine is determined by titration using thiosulfate with starch as indicator. Since HI does not react with compounds that cause off-flavors or polymers formed by oxidation, the PV may underestimate the degree of oxidation. Generally, the PV is plotted versus storage time in order to establish the onset of oxidation. The method is sensitive to temperature, and results can be imprecise.

The resistance of lipids to oxidation is important for most products. During accelerated shelf-life tests, oxidation can be artificially increased by raising the reaction temperature, ambient oxygen concentration, and by addition of metal ion catalysts. The time required to the start of oxidation is then measured by the TBA or PV test. In the **active oxygen** method, purified air is bubbled through a sample at a temperature of 110 °C. The amount of formic acid (a volatile acid) formed is then determined. Alternatively, changes in conductivity can be determined continuously. In this way, we obtain a measure of unsaturation and the stability of an oil.[6]

## 5.4.2 IODINE BINDING

The **iodine value (IV)** is a measure of the degree of unsaturation (number of double bonds). The technological significance of this measurement is to characterize and monitor oil hydrogenation. The iodine value is equivalent to the quantity (*in grams*) of iodine adsorbed/100 g sample. The IV is directly proportional to the degree of unsaturation. The procedure for assessing IV is straightforward: (1) the sample of oil is dissolved in a solvent and reacted with a predetermined amount of iodine chloride (or iodine bromide); (2) potassium iodide is added which reduces the excess iodine chloride to free iodine; and (3) the unreacted iodine is then titrated with sodium thiosulfate using a starch indicator.

## 5.5 PHYSICAL PROPERTIES OF TRIACYLGLYCERIDES

Fats and oils have a profound effect on the texture of foods. The delicate mouthfeel of high quality chocolate, the delightful taste of ice cream and the richness of milk cream is partly due to oil and fat droplets. From a materials viewpoint, three physical characteristics are of interest: (1) melting point, (2) polymorphism, and (3) the fat solids index. We now deal with each of these attributes in turn and consider variables affecting them.

### 5.5.1 MELTING POINT

The melting point is the temperature at which a lipid transforms from the solid to the liquid state. The melting temperature for fats is generally broad since fats are a blend of different TAGs. A range of simple techniques is available for assessing melting point:

(1) **Capillary melting point.** The sample of fat is placed in a capillary tube, sealed, and heated slowly in a water bath until completely clear or the solid fats index (SFI) becomes zero.
(2) **Wiley melting point.** A fat sample is formed into a disk and then chilled to solidify. The disk is then placed in a temperature programmed water bath. The melting point is that temperature at which the disk undergoes a loss of shape.
(3) **Dropping melting point.** In this automated (instrumental) method, the sample is placed in a cup that has a hole in the bottom. Upon heating the cup, the fat melts and flows from the cup. The melting point is obviously the temperature at which flow occurs.[6]

The melting point of a fat is determined by the type of fatty acid constituents and their position on the glycerol backbone. Variables that affect the melting point of pure fatty acids probably affect triacylglycerides in a similar way. The following discussion applies to fatty acids and TAGs interchangeably. The melting point of a fatty acid increases with the chain length and with increasing **degrees of saturation** (Table 5.3). **Geometric isomerism** also affects melting, with *trans* fatty acids possessing higher melting points compared to their cis isomers. The order of increasing melting point (saturated fatty acid > *trans* fatty acids > *cis* fatty acids) can be understood based on the model for the melting process described in Chapter 7. **Crystallization** requires the presence of multiple nuclei or tiny accretions of molecules within the oil phase. Each nuclei then grows in size as more molecules adhere to the crystal surface. For a crystal to increase in size the probability of accretion must exceed the probability of disaggregation. Beyond the nucleation process, the rate of crystallization is described by first order kinetics. Over the temperature range 0.5–17°C the activation energy for fat crystallization is negative (−38.5 kJ/mole), which implies that the rate of crystallization increases with decreasing temperature.[7]

During **melting** the ordered, crystalline array of TAG molecules is transformed into a disordered, mobile, ensemble within a liquid state. This requires an input of kinetic energy in the form of heat. The generally low melting point of *cis* fatty acids is explained by the marked "kink" within their structure (Figure 5.4). This irregular shape leads to inefficient packing of *cis* fatty acids as compared with *trans* fatty acids and saturated fatty acids within a crystalline network.

### 5.5.2 POLYMORPHISM

Polymorphism refers to the formation of TAG crystal structures which differ in the arrangement of molecules. **Lateral packing** of fatty acid chains leads to short spacings whilst **longitudinal stacking** of TAG molecules into layers yields long spacings. Information about these two orders of structures can be obtained by X-ray diffraction. Most frequently encountered are three lateral arrangements leading to so-called $\alpha$, $\beta'$, and $\beta$ crystal polymorphs, in order of their increasing heat stability. Other intermediary polymorph states, usually designated as sub-$\alpha$, sub-$\beta'$, $\beta'_1$, $\beta'_2$ etc. have also been reported in the literature. As described above, the long spacings from X-ray diffraction, due to longitudinal staking of layers of TAG molecules, are usually equivalent to **double or triple** the chain length of fatty acids.[8]

### 5.5.2.1  Polymorphism of Single Triacylglycerides

Clarkson and Malkin (1934)[9] showed that tristearin (SSS) formed four crystal structures ($\gamma$, $\alpha$, $\beta'$, and $\beta$) as judged from X-ray diffraction and melting point ($T_m$) results. The $\gamma$ polymorph was subsequently contested. The present view is that there are three major polymorphs.[7] The crystallization and melting behaviors of a symmetrical TAG, which contain only one type of fatty acid can be summarized as shown in Eqs 5.2–5.4:

$$\text{oil} \rightleftharpoons \alpha \rightarrow \beta' \rightarrow \beta, \tag{5.2}$$

$$\beta' \rightarrow \text{oil}, \tag{5.3}$$

$$\beta \rightarrow \text{oil}. \tag{5.4}$$

These equations show that during rapid cooling a TAG is transformed from a liquid (oil) state to the $\alpha$ state. In this $\alpha$-polymorphic state, fatty acid chain ($CH_2$–$CH_2$) segments retain a high degree of mobility and are arranged in a disordered or hexgonal packing array. The $\alpha$ state has a lifetime of about 60 s after which it is converted to the $\beta'$-**orthorhombic** packing arrangement (lifetime 60 s to years). The final polymorphic state, called the $\beta$ **parallel** or **triclinic** packing arrangment, is the most stable. Both $\alpha$ and $\beta'$ structures are metastable (unstable) in that they eventually transform into the $\beta$ crystal (Eq. 2). Apparently, TAGs are liable to exist as $\alpha$ crystals during processing but transform into the $\beta'$ crystalline state during storage.[10] A representation of the subcell structure is shown in Figure 5.5.

More complex crystallization behavior can be expected when a TAG contains two or more different fatty acids. Sato et al.[11] and Koyano et al.[12] reported five and six crystal forms for SOS and POP.[a] Four of the crystal states ($\alpha$, $\gamma$, $\beta_1$, $\beta_2$) were identical for POP and SOS (Table 5.4). In addition, POP appears to produce three further crystalline forms ($\delta$, pseudo-$\beta'_2$, pseudo-$\beta'_1$). The $\gamma$ form for POP is observed with samples of 99.2% purity but not with 99.9%

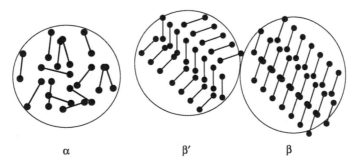

$\alpha$ $\qquad\qquad\qquad$ $\beta'$ $\qquad\qquad\qquad$ $\beta$

**Figure 5.5** Diagrammatic representation of the subcell structure for different triacylglyceride polymorphic states. The alpha form has a hexagonal subcell structure with a Bragg spacing of 0.42 nm. The $\beta'$ form has an orthorhombic perpendicular subcell packing with a Bragg spacing of 0.42–0.43 nm. The $\beta$-form has a triclinic parallel subcell structure with a Bragg spacing of 0.46 nm.

---

[a]For convenience TAGs are described using single letter abbreviations for fatty acids (Table 5.3); SOS=1,3 di-steroyl-2-oleoyl glycerol and POP=1,3-di-palmitoyl-2-oleoyl glycerol.

**Table 5.4**
**Melting characteristics of polymorphs of SOS and POP**

| Polymorph | $T_m$(°C) | | $\Delta H_f$ (kJ/mol) | |
|---|---|---|---|---|
| | SOS | POP | SOS | POP |
| Alpha ($\alpha$) | 23.5 | 15.2 | 48 | 68 |
| Gamma ($\gamma$) | 35.4 | 27 | 99 | 92 |
| Pseudo-$\beta'$ | 36.5 | – | 105 | – |
| $\delta$ | – | 29.2 | – | 107 |
| Pseudo-$\beta'_2$ | – | 30.3 | – | 95.5 |
| Pseudo-$\beta'_1$ | – | 33.5 | – | 98.3 |
| $\beta_2$ | 41.0 | 35.1 | 143 | 124 |
| $\beta_1$ | 43.0 | 36.7 | 151 | 130 |

From Sato et al. (1989).[11] $T_m$ = melting temperature and $\Delta H_f$ = heat of formation for different polymorphs. The entropy of formation $\Delta S_f$ (not shown) = $\Delta H_f / T_m$ (K). The precision of this data is ±1°C and ±3–7 kJ/mol for $T_m$ and $\Delta H$ measurements.

pure samples. These results were confirmed in a number of studies involving differential scanning calorimetry and Fourier transform infrared analysis.[13] Melting characteristics of polymorphs of SOS and POP are given in Table 5.4.

The melting temperature and heats of formation increase from the less stable to the more stable crystalline state (Table 5.4). Both parameters reflect the overall **cohesive energy** within different polymorphic crystals states arising from lateral chain–chain interactions as well as interactions between different layers of a TAG. From thermodynamic considerations the more ordered $\beta_1$ crystalline structure must eventually form.

The **kinetics** of crystallization was monitored using a polarizing optical microscope fitted with a photosensor.[12] The sample of oil to be studied is placed on a thermostatted microscope stage connected to a water bath. The amount of birefringent light passed is then measured in terms of millivolt output from the photosensor, as the sample of oil is heated or cooled. The crystallization rates and temperatures for SOS are summarized in Figure 5.6. The results show that less stable SOS polymorphs formed rapidly whilst more highly ordered TAG crystal structures formed slowly. Note that the transitions taking place at lower temperatures would be more rapid than those at high temperatures owing to the negative activation energy for crystallization. Transformations of crystal polymorphs for SOS are shown in Figure 5.7.

### 5.5.2.2 Polymorphism in Natural Edible Oils

Crystallization is relatively simple for so-called symmetrical TAGs containing only one type of fatty acid. When a TAG has more than one type of fatty acid an extra level of complexity arises because crystallization has to accommodate the differently sized fatty acids. We now consider a third level of complexity expected for natural oils which are mixtures of TAG molecules. Cocoa butter contains three classes of TAG: saturated (3%), polyunsaturated (13%), and monounsaturated (80%). The monounsaturated fatty acid fraction comprises SOS (25%), POS (50%), and POP (25%).

The crystallization behavior of cocoa butter has been studied extensively since the 1960s. The goal of these studies was to improve our understanding of chocolate and help to improve product quality and keeping characteristics. These studies involved X-ray diffraction analysis. The description of cocoa butter crystallization provided by Wille and Lutton (1966),[14] and later confirmed by others[15] is well accepted. Commercial cocoa butter from the Hershey Chocolate Corporation was pretreated by mixing with a 3% slurry of kieselguhr and bleaching earth to remove suspended solids. The samples were then subjected to melting point and

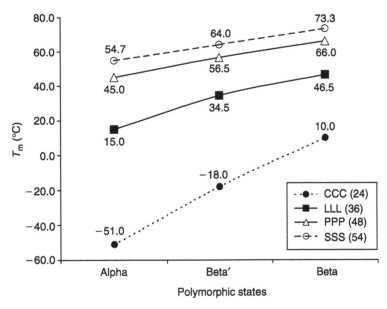

**Figure 5.6** The melting temperature ($T_m$) for $\alpha$, $\beta'$, and $\beta$ polymorphic states for symmetrical triacylglyerides. Boxed legend shows the total number of carbon atoms for tricraprylin (CCC), trilaurin (LLL), tripalmitin (PPP) and tristrearin (SSS). Drawn from data in Guillen-Sans and Guzman-Chozas.[4]

Dù J (15 °C, 1.2 h),

Jù J pseudo-E (35 °C, 1.5 h),

pseudo-Eù $E_2$ (35 °C, 8 h),

$E_2$ ù $E_1$ (40 °C, 2.5 days)

**Figure 5.7** Transformations of crystal polymorphs for SOS.

X-ray diffraction analysis. Further investigations were done using single TAGs as well as binary and ternary mixtures of POP, SOS, and POS. The results are summarized in Table 5.5.

Initial studies of the crystallization behavior of POP, POS, and SOS by themselves showed that each TAG formed four to five polymorphs. More recent studies indicate that additional polymorphs are formed by POP, POS, and SOS. Cocoa butter itself forms six polymorphic states designated I–VI. Each crystalline state appears to be a combination of the polymorphs formed by the pure TAGs. Using contemporary notation cocoa butter states I–VI correspond to six structures in order of their increasing stability: $\beta'$sub($\alpha$), $\alpha$, $\beta'_2$, $\beta'_1$, $\beta_2$, $\beta_1$. The lamellae size for states I–IV (cf. long spacing) equals two fatty acid chain lengths whilst states V and VI have a three fatty acid repeated long spacing corresponding to a parallel, triclinic arrangement within the unit cell.

The number of polymorphs formed by a given TAG is usually uncertain. Crystallization seems an inherently temperamental process and sensitive to the presence of impurities (as discussed above). Other variables such as sample thermal history, cooling rate, presence of nuclei (nature of seeding if any), degree of mechanical agitation, and TAG composition also affect crystallization. Research for cocoa butter indicates that crystallization follows nucleation of a high melting ($T_m = 72°C$) minor component that has unusually high levels of phospholipids and glycolipids.[16]

**Table 5.5**
**Polymorphs for TAG and the I–VI states for cocoa butter**

Polymorphs for single TAG

| | |
|---|---|
| SOS | $\beta_3$, $\beta_{3sub}$, $\beta'_3$, $\alpha_3$, $\alpha_{3sub}$ |
| POS | $\beta_3$, $\beta_{3sub}$, $\beta'_3$, $\alpha_3$, $\alpha_{3sub}$ |
| POP | $\beta_3$, $\beta'_2$, $\beta'_{2sub}$, $\alpha_2$, $\alpha_{3sub}$ |

| Polymorphs of cocoa butter | | $T_m$ (°C) | Storage stability |
|---|---|---|---|
| VI | $\beta_3$ | 36.3 | 26°C, stable |
| V | $\beta_{3sub}$ | 33.8 | 16°C, 2–14 wks |
| IV | $\beta_3$, $\beta'_2$, $\beta'_{2sub}$ | 27.5 | 16°C, 2 days to 2 wks |
| III | II + IV | 25.5 | 10°C, 1–3 days |
| II | $\alpha_2$, $\alpha_{3sub}$ | 23.3 | 5°C, < 2 h |
| I | II + another | 17.3 | 0°C, 15 min |

## 5.6 FUNCTIONALITY OF FATS AND OILS

Fats impart physical properties to foods and thereby affect the sensory, nutritional, safety, and storage characteristics. Much work has focussed on the texture of fats as ingredients. The goal of this research is to explain macroscopic properties in terms of molecular structure, melting, and crystallization behavior.

The **viscosity** of edible oils is possibly the result of interactions between TAG molecules probably via Van der Waals' interactions. Experience with alkanes suggests that viscosity will increase with increasing chain length. However, a more complete discussion of viscosity must incorporate the melting characteristics.

The **texture** of a TAG depends on several physical attributes, chief amongst which are the **melting properties**, the **solid fat content**, and **polymorphism**. Crystallization behavior is affected by the presence of different fatty acid substituents on the glycerol backbone and the number of TAG types present in different edible oils. When an oil solidifies, TAG molecules having "compatible" fatty acids may tend to co-crystallize. This leads to the formation of fat crystal particles or crystallites suspended in a melted oil continuous phase. Interactions between dispersed particles leads to a three-dimensional **fat crystal network** entrapping the liqiud oil phase.[17,18]

### 5.6.1 PROPERTIES VERSUS QUALITIES OF FATS

A distinction is necessary between the **physical properties** of fats and oils and their **quality**.[19] The complicated relations between quantitative instrumental measurements and sensory perception by human subjects were discussed earlier. Physical characteristics such as viscosity, melting point, and solid fat content can be readily measured quantitatively. In contrast, subjective quality attributes are usually measured via sensory analysis using trained panelists. Use of these different analytical approaches is essential because quality is a **multidimensional** as well as a **subjective** attribute. Quality is determined by a combination of elementary physical characteristics. The subjective element arises because human perception has **psychological** and **physiological** aspects and because there are differences between subjects. The following quality attributes for oils and fats depend on their **viscoelastic** behavior: texture (solids), consistency (semi-solids), and viscosity (liquids).

Fat **lubricity** is a quality of reducing friction linked with the perception of creaminess and smoothness. Waxiness arises from the tendency for high melting fats to coat the oral cavity. Solid fat content and crystal size together determine the quality of **graininess** and **hardness**. The **consistency** of fats is associated with the balance of liquid–solid character or viscoelasticity.

**Table 5.6**
**Functionality of fats and oils**

| Attribute | Elementary physical property | Quality |
|---|---|---|
| Texture | Viscosity | Consistency |
|  | Plasticity, melting | Cohesiveness |
|  | Shortening | Spreadability, mouthfeel |
| Surface activity | HLB number | Lubricity |
|  | Lubricity |  |
| Flavor | Flavor carrier, flavor binding | Aroma |
| Stability | Chemical and enzymatic oxidation | Odors, keeping quality |

**High fat foods**
Butter and spreads
Bread, cakes, cookie doughs,
   and butters
Cooking oil, salad oil, and dressings
Confectionery
Cultured dairy products (cheese,
   yogurt, sour cream)
Frozen deserts (ice cream, whipped
   topping)

Creamy, stiff, smooth, chewy, firm, gummy, and gluey are further qualitative descriptors for consistency. **Mouthfeel** can be correlated with zero-shear viscosity ($\eta_0$) and shear rate needed to reduce this value by 90% (Chapter 8, Section 8.6.6). However, mouthfeel is perhaps the final integrand: this attribute therefore is dependent on all preceding functionalities. A summary of fat functionalities and of foods where these are important is shown in Table 5.6.

## 5.6.2 PLASTICITY

A plastic fat flows when force is applied and regains its original *consistency* (not shape) when the force is removed. Another feature of a plastic fat (margarine, coconut oil, palm oil) is its tendency to melt over a wide temperature range. Plasticity is ascribed to the presence of a mixture of crystal polymorphs having different stability. Nonplastic fats (cocoa butter) produce a sharp melting transition from solid to liquid when heated over a range of temperatures. The melting profile of plastic fats is broad whereas nonplastic fats exhibit sharp melting profiles. Plasticity also refers to the property of a material to undergo permanent deformation. For edible fats, this quality translates into **spreadability**.[20] According to rheological terminology, plastic fats have a low yield value and show non-Newtonian behavior.

    **Solid fat content** and **crystal particle size** affect plastic behavior. For instance, the penetration hardness of spreadable fats was proportional to the second power of the solid fat content. Turning to the relation between particle size and the rheology of plastic fats, the general finding is that fine grained fats tend to be harder than coarse crystal fats. Viscoelastic studies indicate that shear modulus may be proportional to the third power of particle diameter.[17]

    The plasticity of **milk fat** has received much attention owing to its role as **butter**. At room temperature, butter must remain a free standing solid and be spreadable with a slight application of force. The spreadability of butter is associated with a broad melting temperature range over which both solid and liquid TAG fractions co-exist. Plasticity is the result of an interacting 3-dimensional network of fat crystals within a continous oil phase. Plasticity is affected by TAG composition. The long chain saturated fatty acids fraction is necessary for the plastic behavior associated with butter. Removal of saturated fatty acids (~20% total) from milk produces a pourable oil at room temperature.

### 5.6.3  SHORTENING AND ANTI-STALING

Shortening produces a **tenderizing action** on baked cereal products. This effect arises when protein particles become coated with fats, leading to inteference with the **gluten network** formation. Fats and oils, as well as the breakdown products (free fatty acids, mono- and di-acylglycerides), interact with carbohydrates. Formation of the amylose–fatty acid complex appears to affect cold hardening of baked bread—staling. These interactions contribute to the quality of moistness and help to maintain soft crumb texture.

### 5.6.4  FAT MICROSTRUCTURE AND CRYSTAL NETWORKS

In the 1990s researchers in the United States started to examine ways to explain the effect of crystal structure on the mechanical and rheological properties of fats.[21] The problem was how to explain the **macroscopic** properties of fats in terms of fundamental chemical properties or **microstructure** (Figure 5.8). According to Narine and Marangoni, knowledge of TAG polymorph ($\alpha$, $\beta'$, and $\beta$) subcell structure is based on a limited number of symmetrical TAGs. These investigators also doubted whether techniques for studying subcell structure (X-ray diffraction) could provide information about fat microstructure upon which macroscopic properties could be predicted.

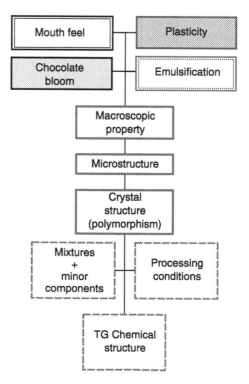

**Figure 5.8** Variables affecting fat macroscopic properties. Fractal mathematics links physical microstructure and functional properties such as viscoelasticity. Modified from Narine and Marangoni.[21]

TAG molecules are arranged within different crystalline arrays at a molecular spacing of about $10^{-9}$ m. By contrast, fat microstructural elements are arranged within a spacing of $10^{-6}$ m to $10^{-4}$ m. Our perceptual discrimination during eating and mechanical treatments covers the larger size range.

**Fractal geometry** was employed to account for the power dependence of elastic modulus ($G'$) on particle volume fraction ($\Phi$):

$$G' \sim \Lambda \Phi^{m} \tag{5.5}$$

where $\Lambda$ is a constant that depends on the size of the microstructural elements under shear and $m$ is a fractional number. Moreover, fat crystal microstructure could be described using **fractals**. This approach allows a geometric description of objects that possess nonintegral dimensions. For instance, 1-dimensional, 2-dimensional, and 3-dimensional objects correspond to a line, area, and volume, respectively. Fractal objects possess *fractional* dimensions. Fractal objects are geometrically **self-similar** and usually exhibit **self-repeating** architecture spanning orders of magnitude of scale.

Narine and Marangoni suggested that fractals may provide the conceptual framework for linking (**scaling up**) hierarchical orders of structure and functionality from the molecular to microstructural level. The repeating dendritic structure of trees provides a well known example of fractal structure. The overall shape of a tree appears to be built from a self-repeating pattern evident at the scale of the single branch right up to the whole tree. Fat microstructure, and the dependence of rheological properties on $\Phi$ and solid fat content could be explained using fractal mathematics. In Eq. 5.5, the power term ($m$) has a **fractional value**. The results point to the importance of microstructural elements in relation to physical functionality of fats. Fractal mathematics has also been applied to protein functionality.

## 5.7 FAT REPLACERS AND MIMETICS

There are nearly 300 materials proposed as fat replacers. They include modified starches, fiber, gums, emulsifiers, restructured protein, and cellulose. That so many materials can act as fat replacers speaks for the many roles played by fats as ingredients in foods. This section provides a brief overview of the **fat replacers** which are either fat substitutes or mimetics. **Fat substitutes** are lipid-like substances intended to replace fats on a one-to-one basis. **Fat mimetics** are protein or carbohydrate ingredients which function by imitating the physical, textural, mouthfeel, and organoleptic properties of real fats (Figure 5.9).

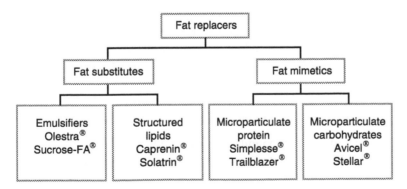

**Figure 5.9**  Classification of the fat replacers as substitutes or mimetics.

### 5.7.1 RATIONALE OF FAT REPLACERS

**Fat replacers** were developed in response to consumer concern with quantities of dietary lipid and its possible link with obesity. In this section we describe the main groups of fat replacers and their functional role in foods.

### 5.7.2 FAT SUBSTITUTES

The challenge for food technologists is to maintain optimal food texture and organoleptic properties whilst reducing fat content. The **fat substitutes** are intended to replace the functionality of natural fats on a one-to-one weight basis. The functionality of natural fats is far from understood (Section 5.6). The successful adoption of replacers should help define fat functionality. It is relatively easy to duplicate **emulsification** using nondigestible surfactants in place of the surface active natural lipids such as **phospholipids** (lecithin), **monoglycerides**, and **diglycerides**.

Olestra®, also known as Olean, is an emulsifier produced by reacting sucrose with 6–8 moles of C12–C22 fatty acids in the presence of a catalyst. Olestra was approved in 1996 for use in savoury foods. Less completely esterified sucrose fatty acid esters (2–3 moles of fatty acids) have been produced which are hydrophobic and also more digestible. Similar surfactants have been produced by attaching a hydrophobic fatty acid to hydrophilic sugar alcohols: Sobestrin is produced by esterifying sorbitol. Other polyols used for sugar–fatty acid esters include tetrahalose, raffinose, and stachyose.

The **medium chain triacylglycerides** (MCTs) differ from natural fats by their relative absence of long chain fatty acids. To produce MCTs ordinary vegetable oil is first hydrolyzed and medium and short chain saturated fatty acids are isolated by fractionation. The medium fatty acid fraction is then reattached to glycerol with the aid of a catalyst. Compared to ordinary fats, MCTs have lower melting points, high solubility in water, and are resistant to oxidation. The MCTs are not transferred to the body's store of adipose tissue but are metabolized directly in the liver.

A well known example of an MCT is **Caprenin**®, in which a glycerol backbone is substituted by caprylic acid (C8:0), capric acid (C10:0), and behemic acid (C22:0). This fat replacer yields 5 kcal/g as compared to 9 kg/g for normal fat. The caloric reduction is apparently due to the less efficient metabolism of C8:0 and C10:0 fatty acids within the body[b]. The functional properties of Caprenin are similar to those of butter. The usable temperature for Caprenin is $< 132°C$. Examples of fat substitutes and fat mimetics are given in Tables 5.7 and 5.8, respectively.

**Salatrim** (<u>s</u>hort <u>a</u>nd <u>l</u>ong <u>a</u>cyl <u>t</u>riglyceride <u>r</u>earranged <u>m</u>olecules) is another structured lipid produced from a mixture of short chain (C2:0–C4:0) and a long chain fatty acid (C18:0). The short chain acids are esterified at position 1,3 of the glycerol molecule whereas the long chain acid is esterified at position 2. Salatrim, also called Benefat™, provides approximate 5 kcal/g because the short acids provide few calories whilst stearic acid is only partically absorbed in the body. Salatrim is usable at pH 3–7.5 and under cold conditions. Suitable product categories for salatrim include confectionery, cookies, cakes, brownies, and pie crust.

### 5.7.3 FAT MIMETICS

#### 5.7.3.1 Carbohydrate Based Fat Mimetics

The fat mimetics are nonlipid compounds that are able to simulate the physical functionalities of fats, namely qualities like creaminess and smoothness.[22] **Carbohydrate** fat mimetics (Avicel®

---

[b]In contrast one should expect the heat of combustion for different fats to be the same.

**Table 5.7**
**Examples of fat substitutes***

| Name/group | Composition |
| --- | --- |
| **Emulsifiers** | |
| Olestra®, Olean® | Sucrose esterified with 6–8 fatty acids |
| Sucrose fatty acid ester | Sucrose esterified with 1–3 fatty acids |
| Sorbestrin® | Carbohydrate + fatty acids |
| Sorbitol–fatty acid ester, | |
| Alkyl glycoside polyester | Alkyl glycoside + fatty acid |
| Span® and Tween® | Sorbitan monostearate (Spans), |
| Miscellaneous | Polyoxyethylene fatty acid esters (Tweens) |
| | Polypropylene fatty acid esters |
| **Structured lipids** | |
| Caprenin® | Medium chain trycerides (MCT) |
| | C8:0, C10:0, C22:0 |
| Neobee® | Medium chain trycerides (MCT) from coconut oil, palm oil |
| Salatrim/Benefat® | MCT; C2:0 to C4:0, C18:0 fatty acids |
| Dialkyl dihexdecyl molanate | |

*Details for Tables 5.7 and 5.8 were adapted from various sources including the International Food Information Council (IFIC) http://www.ific.org

**Table 5.8**
**Examples of fat mimetics**

| Name/group | Composition |
| --- | --- |
| **Microparticulate protein** | |
| Simplesse® | Microparticulate whey protein, egg protein |
| Trailblazer® | Microparticulate egg protein, milk protein |
| Dairylo® | Whey protein |
| Lita® | Corn gluten |
| **Microstructure carbohydrates and gelling agents** | |
| Avicel® | Cellulose microparticles |
| Methocel® | Cellulose ethers |
| Amalean® I & II, Farinex™, | Modified starches |
| Instant Stellar™, Perfectamyl™ AC, | |
| PURE-GEL®, STA-SLIM™ | |
| Kel-Lite KELCOGEL®, KELTROL®, Slendid™ | Xanthum, pectin, alginates, and other structural polysaccharides |
| Oatrim® | Oat derived, forms thermoreversible gels |
| Slendid | Microparticulated pectin gel particles |

cellulose gel, Methocel™, Solka-Foc™) include microparticulate cellulose. These materials provide the mouthfeel and flow properties of fat but are lacking the flavor characteristics associated with edible fats. Microparticulate cellulose also **retains moisture** and acts as a **texturizer** and **stabilizer**. Carbohydrate fat mimetics have been suggested as useful in dairy products, sauces, frozen desserts, and salad dressings. The FDA (USA) classifies carbohydrate fat mimetics as GRAS (generally regarded as safe) ingredients.

The second class of carbohydrate based fat mimetics includes starch and **modified starches** from potato, corn, oat, rice, wheat or tapioca. Their main functions are as bodying agents and texture modifiers intended to be used with emulsifiers, proteins, gums, and other modified food starches. The modified starches have FDA status as approved food additives and may be found in processed meats, salad dressings, baked goods, fillings and frostings, sauces, condiments, frozen desserts, and dairy products.

**Oatrim** (Beta-Trim™, TrimChoice) produces 1–4 kcal/g. This type of fat mimetic is made by partial enzymatic hydrolysis of oat starch. The water soluble products of oat flour are used to replace fat and as a texturizing ingredient. Oatrim is classed as GRAS by the FDA. This ingredient may be used for baked goods, fillings and frostings, frozen desserts, dairy beverages, cheese, salad dressings, processed meats, and confections.

USDA researchers developed **Z-trim** from insoluble fiber from oat, soybean, pea, and rice hulls or from corn or wheat bran. The mouthfeel of Z-trim is similar to fat in terms of its moistness, density, and smoothness. Z-trim has the further advantages of increasing the fiber content of food. This ingredient has FDA status and can be used with baked goods, burgers, hot dogs, cheese, ice cream, and yogurt. Although heat stable, Z-trim is not considered suitable for frying.

### 5.7.3.2  Protein Based Fat Mimetics

The microparticulated proteins include Simplesse® (1–4 kcal/g) which is made from **whey protein** or milk and egg protein. The manufacturing process involves simultaneous heating and shearing to produce small particles of coagulated protein. This material provides the mouthfeel of fat. Like most of the other fat mimetics Simplesse is not suitable for frying but is stable for baking. FDA has granted Simplesse GRAS status. The ingredient may be used in dairy products, salad dressing, maragarine- and mayonnaise-type products, baked goods, coffee creamer, soups, and sauces.

Dairy-Lo® (4 kcal/g) is a protein based ingredient manufactured via thermal denaturation of proteins from sweet whey. It improves the texture, flavor, and stability of low-fat foods. Dairy-Lo® provides the mouthfeel of fat. It has GRAS status and is considered suitable for use in milk and dairy products, baked goods, frosting, salad dressing, and mayonnaise-type products. Dairy-Lo is not considered suitable for frying.

## References

1. Becker, W., Solvent extraction of soybeans, *J. Am. Oil Chem. Soc.* 55(Nov), 754–761, 1978.
2. Nawar, W. W., Lipids, in *Food Chemistry* (3rd edition), Fennema, O. R., Ed., 225–319, 1996.
3. Firestone, D. and Sheppard, A., Determination of trans fatty acids, in *Advances in Lipid Methodology*, P. J. Barnes & Assoc, Bridgewater, UK, 1992.
4. Guillen-Sans, R. and Guzman-Chozas, M., The thiobarbituric acid (TBA) reaction in foods: a review, *Crit. Rev. Food Sci. Nutr.* 38(4), 315–330, 1998.
5. Fernandez, J., Perez-Alvarez, J. A., and Fernandez-Lopez, J. A., Thiobarbituric acid test for monitoring lipid oxidation in meat, *Food Chem.* 59(3), 345–353, 1997.
6. Firestone, D., Ed., *Official Methods and Recommended Practices of the American Oil Chemists Society*, 4th edition, AOCS, Champaign, IL, 1990.
7. Mortensen, B. K., Physical properties and modification of milk fat, in *Developments in Dairy Chemistry—2*, Fox, P. F., Ed., Applied Science Publishers, New York, 159–194, 1983.
8. Garti, N. and Sato, K., *Crystallization and Polymorphism of Fats and Fatty Acids* (Surfactant Science Series, Vol. 31), Marcel Dekker Inc., New York, 1988.
9. Clarkson, C. E. and Malkin, T., Alternation in long chain compounds. Part II. An X-ray and thermal investigation of the triglycerides, *J. Chem. Soc.* 666–671, 1934.

10. DeMan, J. F., *Principles of Food Chemistry*, Aspen Publishers, Gaithersberg, MD, 33–110, 1999.

11. Sato, K., Arishima, T., Wang, Z. H., Ojima, K., Sagi, N., and Mori, H., Polymorphism of POP and SOS. I. Occurrence and polymorphic transformation, *J. Am. Oil Chem. Soc.* 66(5), 664–674, 1989.

12. Koyano, T., Hachiya, I., Arishima, T., Sato, K., and Sagi, N., Polymorphism of POP an SOS. II. Kinetics of melt crystallization, *J. Am. Oil Chem. Soc.* 66(5), 675–679, 1989.

13. Yano, J. and Sato, K., FT-IR studies on polymorphism of fats. Molecular structures and interactions, *Food Res. Intern.* 32(4), 249–259, 1999.

14. Wille, R. L. and Lutton, E. S., Polymorphism of cocoa butter, *J. Am. Oil Chem. Soc.* 43, 491–496, 1996.

15. Loisel, C., Keller, G., Lecq, G., Bourgauz, C., and Ollivon, M., Phase transitions and polymorphism of cocoa butter, *J. Am. Oil Chem. Soc.* 75(4), 425–439, 1998.

16. Davis, T. R. and Dimick, P. S., Lipid composition of high-melting seed crystals formed during cocoa butter solidification, *J. Am. Oil Chem. Soc.* 66, 1494–1498, 1989.

17. De Man, J. M. and Beers, A. M., Fat crystal networks: structure and rheological properties, *J. Texture Studies* 18, 303–318, 1987.

18. Juriaanse, A. C. and Heertje, I., Microstructure of shortenings, margarine and butter: A review, *Food Microstructure* 7, 181–188, 1988.

19. Schaich, K. M., Rethinking low-fat formulations—Matching fat functionality to molecular characteristics, *The Manufacturing Confectioner*, June, 109–122, 1997.

20. German, J. B. and Dillard, C. J., Fractionated milk fat: Composition, structure and functional properties, *Food Technol.* 52(2), 33–38, 1998.

21. Narine, S. S. and Marangoni, A. G., Relating structure of fat networks to mechanical properties: a review, *Food Res. Intern.* 32, 227–248, 1999.

22. Fat Reduction in Foods, White Paper from the Calorie Control Council, Atlanta, GA, August, 1996.

# 6

# Proteins

## 6.1 INTRODUCTION

A great deal is known about proteins and their role in biology and medicine. The shapes of proteins are linked to their biological function. **Fibrous proteins** have an elongated shape and are insoluble. These characteristics are important for their role as structural components; for example, connective tissue and muscle (Table 6.1). By comparison, **globular** proteins are compactly folded, amphiphilic, soluble, and suited for their role as ligand binders or enzymes. Seed **storage proteins** have repeating **block sequences** of amino acids. This **structural motif** allows proteins to be packed in a semi-dehydrated state within plant seeds. Caseins possess an extended structure that makes them highly digestible by young mammals. The structural elements needed for protein functionality as gelling or foaming agents are not immediately clear. The relation between protein structure and functionality in food is slowly emerging.

The characteristics of **high-protein foods** (meat, fish, eggs, milk, cheese, bread, cakes) depend on the types and amounts of proteins present. Of the four macroconstituents (fats, carbohydrates, water, and proteins) proteins exhibit the widest range of functionality. Protein **ingredients** are used in food formulations to achieve texturization through emulsification, gelation, foaming, and water holding. Polysaccharides can also form gels but are relatively poor **emulsifying** agents. Lipids can serve as good emulsifying agents (lecithin, mono- and diglyercides) but are relatively weak gelling agents. The versatility of protein ingredients stems from their complex structure.

### 6.1.1 PROTEIN SCIENCE AND TECHNOLOGY

Each protein has a **specific 3-dimensional shape** connected to its biological function. Fundamental studies over the past 50 years have revealed some the rules for **protein folding**. There are several *trillion* shapes available to any sizeable protein. Somehow the protein chain searches through the vast number of possibilities and adopts a functioning shape. An amazing feature of protein folding is the speed at which it occurs. A moderately sized protein such as hemoglobin folds within a time frame of milliseconds. If this process were random, folding would occur over periods of years, and life as we know it could not have developed. Protein **misfolding** does occur, leading to dysfunctional structures associated with degenerative diseases like Alzheimer's disease, arthritis, Creutzfeld–Jacob disease, and some forms of cancer. A great deal of protein research is driven by medicine and pharmaceutical applications. The **human genome** project mapped the complete genetic information in a human cell. Emphasis is now shifting to **proteomics** which aims to understand sequential protein expression during different stages of the cell's life cycle. Techniques are being deployed for wholesale identification and quantification of tissue proteins as a function of time, tissue type, and environmental factors. The fields of **proteomics, genomics,** and **bioinformatics** are at the forefront of protein research.

**Table 6.1**

**The roles of proteins in living cells**

- Outer covering in animals (skin, hide, hair, nails, hoofs, and claws)
- Structure and movement (bones, muscles, and tendons)
- Transport of nutrients (blood and components, e.g., hemoglobin, plasma proteins)
- Nutrition of the young (milk, eggs, seeds, and nuts)
- Catalysts (enzymes)
- Defense proteins (antibodies, enzyme inhibitors, lectins, and toxins)

**Table 6.2**

**Some major food protein ingredients**

| Protein | Sources | Functionality/uses |
|---|---|---|
| Dried non-fat milk, whey, protein concentrate, caseinate | Milk | Adhesion, emulsification, gelation |
| Collagen, gelatin, myosin, blood | Meat, fish, and poultry | Gelation, foaming, adhesion |
| Soybean flour, concentrate, isolates from wheat (gluten), maize | Cereals, legumes, oilseeds | Viscoelasticity |
| Liquid or powdered egg albumin, yolk | Eggs | Foaming, whipping agents |

**Protein engineering** is another technology to arise from recent advances in genetics and biotechnology. New genes can be introduced into cells using plasmids as vehicles. A cell's protein synthesizing machinery then reads and translates the plasmid based information forming a new protein. This **genetic engineering** allows the production of modified proteins with improved characteristics, and of novel proteins that would not otherwise occur naturally in a given organism or tissue. Proteins can also be modified by chemical or physical means.

### 6.1.2 FOOD PROTEINS

Any protein ingested intentionally or non-intentionally can be considered a food protein. The scope of food protein chemistry covers agricultural production, processing, packaging, storage, distribution, retail, and protein function as ingredients. Protein **quality** is an important consideration to the consumer as is **allergy** and other adverse reactions to some proteins. The major protein ingredients, material having 5–10% or more protein, are listed in Table 6.2.

The manufacture of protein ingredients involves standard unit operations including separation, concentration, drying, particle size reduction, and packaging. Example methods are described later. The final products (with the exception of liquid eggs) are usually sold in powder form for easier distribution and retail. Quality control checks are necessary to ensure product consistency, high nutritional quality, low microbiological count, bland flavor, and extended storage life. Commercial protein ingredients are used for large-scale food manufacturing and processing, **catering** or **domestic cooking**. Another outlet for high protein foods is as livestock feed. Some nonfood applications of proteins are in the manufacture of adhesives, items of clothing, and biodegradable packaging.

## 6.2 PROTEIN CLASSIFICATION

Proteins are catagorized according to source, solubility, shape, physiological role, and secondary structure. We might easily think of other categories such as acidic, basic, nonpolar, complex proteins (having nonprotein prosthetic groups), metallo-proteins (proteins with tightly bound metal ions), and glycoproteins (proteins with attached carbohydrate groups). Details of some different protein categories important to food technologists are described below.

### 6.2.1 CLASSIFICATION OF PROTEINS BY SOURCE AND SHAPE

The major sources of dietary protein in the US are listed in Table 6.3. Proteins from animal sources (meat, fish, poultry, eggs) constitute about 70% of the dietary intake. Plant proteins make up another 17%. New and emerging sources of dietary protein include leaf and microbial proteins. The data in Table 6.3 are fairly representative of protein consumption in most industrialized countries. In developing countries, the proportion of animal protein in the human diet is lower and the proportion of plant protein higher than shown in Table 6.3. There are three classes of protein according to their shape. The **globular proteins** are spherical. **Fibrous proteins** have an elongated rope-like shape. The third kind of shape is a **semi-extended** structure found with casein and certain storage proteins in cereal grains.

### 6.2.2 PROTEIN CLASSIFICATION BY SOLUBILITY

Thomas Burr Osborne divided proteins into five classes based on their solubility in a range of solvents (Table 6.4). Sequential extraction of food materials using water, dilute salt solution, and 70% ethanol, followed by dilute alkali recovers five classes of proteins: **albumin** (proteins dissolvable with water), **globulin** (proteins dissolvable with dilute salt), **prolamin** (dissolves with ethanol/water solvent), **glutelin** (dissolves with dilute alkali), plus residue. Details of the Osborne extraction are given in Appendix 6.1. The fractionation is seldom exact. Therefore, some globulins co-extract with albumin and so forth. The Osborne classes of proteins have undergone modification. After examining amino acid sequences for cereal storage proteins,

**Table 6.3**
**Sources of protein in the US diet**

| Protein | % Dietary protein |
| --- | --- |
| Milk, cream, cheese | 22 |
| Meat, fish, and poultry | 43 |
| Eggs | 4.8 |
| Vegetables and fruit | 12.4 |
| Cereals and oil seeds | 17.6 |
| Fats | 0 |
| Sugars and preserves | 0 |

**Table 6.4**
**The Osborne classes of proteins**

| Protein group | Extraction solvent |
| --- | --- |
| Albumin | Water |
| Globulin | Dilute salt (0.1 M sodium chloride) solution |
| Prolamin | 70:30 ethanol:water mixture |
| Glutelin | Dilute alkali (0.1 M NaOH) solution |
| Residue protein | Alkali, urea, and disulfide reducing agents |

Thatham and Shewry concluded these could all be considered sub-types of prolamins. The prolamins subfamilies had differences in sulfur amino acids as well as disulfide bonding. The Osborne nomenclature (Table 6.4) is still widely used in books and journal articles about proteins.

### 6.2.3 CLASSIFICATION AND PHYSIOLOGICAL FUNCTION

Some common functions of proteins in living systems are summarized in Table 6.1. The **fibrous proteins** are relatively insoluble in water, inert, and well suited for their role in building structures like skin, fur, nails, muscle, and tendon. The extended structure of collagen, keratin, and elastin allows strong protein–protein interactions leading to the exclusion of water from the protein matrix. The resulting protein fibrils are further reinforced by multiple interchain bonds. **Globular proteins** (albumins or globulins) are relatively soluble in aqueous solvents, which makes these suited for their role in cell cytoplasm or blood plasma. Flexibility is necessary for binding small ligands or catalysis. Storage proteins (eggs, seeds or milk) have a semi-extended structure that makes them partially insoluble. This allows storage in a semi-concentrated form as protein aggregates or granules. Storage proteins usually need to be mobilized to provide nutrition for the young during germination.

## 6.3 PROTEIN STRUCTURE

### 6.3.1 THE NATIVE CONFORMATION

Each protein has a specific 3-dimensional shape. The different shapes that molecules can adopt without breaking covalent bonds is called conformation. We can grow protein crystals because all proteins of a given type (e.g., beta-lactoglobulin) have essentially the same shape. Proteins that are heterogeneous, due to the presence of carbohydrate components, are difficult to crystallize.

The **native conformation** is that unique 3-dimensional shape which performs a specific biological function. Stabilization of the native conformation and circumstances leading to its loss (denaturation) are discussed in Sections 6.4 and 6.5. In this section, we consider protein conformation and the role of the solvent and how this provides a context for protein folding.

Protein conformation is organized according to hierarchy of structures: primary, secondary, tertiary, and quaternary structure (Table 6.5 and Figure 6.1). According to Anfinsen, protein primary structure contains all information necessary to produce other higher orders of native structure. The primary structure directs the formation of secondary structure that then leads to super-secondary structure, tertiary structure, and higher order structures. This **hierarchical** arrangement of structures may also follow how proteins are manufactured in cells. Beyond the quaternary structure stage, large numbers of proteins interact to produce macroscopic structures ranging from hair, nails, and tendon to muscle.

**Table 6.5**
**Hierarchy of native protein structure**

| Level | Definition |
| --- | --- |
| Primary structure | Sequence of amino acids, patterns of covalent bonding |
| Secondary structure | Localized folding (e.g., $\alpha$-helices and $\beta$-sheets) |
| Super-secondary structure | Association of secondary structure motifs |
| Tertiary structure | Complete folding, overall shape |
| Quaternary structure | Association of protein subunits at specific site |

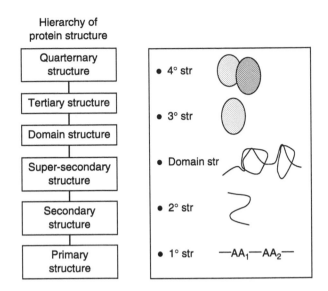

**Figure 6.1** The hierarchy of protein structure.

### 6.3.2 PRIMARY STRUCTURE

Primary structure is the **sequence of amino acids** along a polypeptide chain. Adjacent amino acids are linked by a **peptide bond**. There are an estimated 500 or more amino acids in nature but only 20 alpha amino acids occur naturally in proteins. The number of different amino acids found within food proteins probably exceeds 20 but not by much. Modified amino acids can be produced under physiological conditions (Section 6.3.2.2), via purposeful chemical modification in the laboratory (Section 6.3.2.3) or as a result of food processing (Section 6.3.2.4).

Fibrous proteins have repeating primary structures such as $[A\text{-}B\text{-}C...]_n$. For example, collagen is a regular co-polymer with the formula $(Gly\text{-}X\text{-}Y)_{340}$; position X is usually Pro whilst Y is hydroxy-Pro. Some Lys residues occur in collagen with a low frequency. Wheat storage proteins have a repeat sequence of 7–9 amino acids. Amino acid repeat sequences also occur in the tail region of myosin. These primary **structure repeats** assist in the formation of super-secondary structure (Section 6.3.4).

#### 6.3.2.1 Amino Acid Composition

Alpha amino acids have the general structure shown in Figures 6.2 and 6.3. A central carbon atom ($\alpha$-carbon) is linked to four functional groups: amine ($NH_2$), carboxylic acid (COOH), hydrogen (H), and the **side-chain** (R) group. The 20 naturally occurring side-chain groups in nature account for the 20 different amino acids.

**Amino acid composition** is a simple listing of amino acids found within a protein sample. Unlike primary structure, amino acid composition can be readily altered by changing the blend of proteins in a sample. To determine amino acid composition a known amount of protein is hydrolyzed by heating with 6 M hydrochloric acid at 110°C for 12–72 h. The resulting **hydrolysate** is then separated and quantified using an **amino acid analyzer**. This instrument is essentially a high-performance liquid chromatrophy (HPLC) system fitted with an ion-exchange column or a reverse-phase C18 column. Amino acids eluted from the column can be detected after reacting with a fluorescence compound (*o*-phthalaldehyde) to make them visible in a fluorescence detector. Another common reagent for **post-column derivatization** is ninhydrin. During peptide hydrolysis under acidic conditions, the amino acid tryptophan

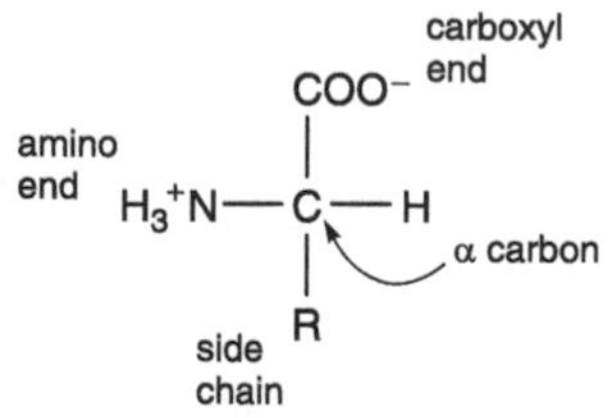

**Figure 6.2** The general structure of an amino acid.

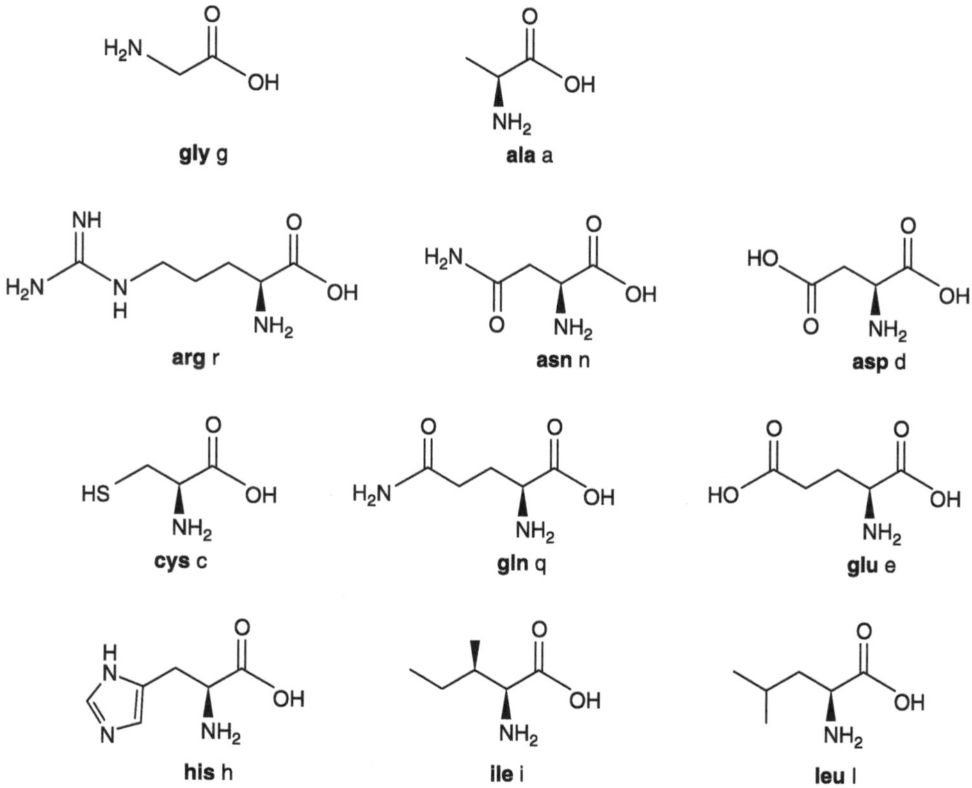

**Figure 6.3** Structures of some amino acids.

is gradually destroyed. Therefore, protein samples intended for tryptophan determination are hydrolyzed by treating with alkaline rather than acid. Structures of amino acids are shown in Table 6.6.

Proteins also undergo hydrolysis during their passage through the stomach and small instestines. About 10–11 naturally occurring amino acids cannot be synthesized by the human body and are classed as **essential amino acids**. For general well being, essential amino acids must be supplied in the diet. Protein **nutritional quality**, which is their ability to support growth, depends on their content of essential amino acids. The currently accepted measurement of protein nutritional quality is **PDCAAS** (protein digestibility corrected amino acid score). The PDCAAS is determined by multiplying two items: (1) percent protein digestibility recorded when a sample of protein is subjected to digestion by a mixture of proteolytic enzymes; and (2) the concentration of the limiting essential amino acid expresssed as a fraction of the recommended daily intake prescribed by WHO for pre-school children.

## Table 6.6
### The structures of amino acids

| Name | Three-letter abbreviation | One-letter abbreviation | Structure |
|------|---------------------------|-------------------------|-----------|
| Alanine | Ala | A | $CH_3-CH(NH_2)-COOH$ |
| Arginine | Arg | R | $HN=C(NH_2)-NH-(CH_2)_3-CH(NH_2)-COOH$ |
| Asparagine | Asn | N | $H_2N-CO-CH_2-CH(NH_2)-COOH$ |
| Aspartic acid | Asp | D | $HOOC-CH_2-CH(NH_2)-COOH$ |
| Cysteine | Cys | C | $HS-CH_2-CH(NH_2)-COOH$ |
| Glutamine | Gln | Q | $H_2N-CO-(CH_2)_2-CH(NH_2)-COOH$ |
| Glutamic acid | Glu | E | $HOOC-(CH_2)_2-CH(NH_2)-COOH$ |
| Glycine | Gly | G | $NH_2-CH_2-COOH$ |
| Histidine | His | H | |
| Isoleucine | Ile | I | $CH_3-CH_2-CH(CH_3)-CH(NH_2)-COOH$ |
| Leucine | Leu | L | $(CH_3)_2-CH-CH_2-CH(NH_2)-COOH$ |
| Lysine | Lys | K | $H_2N-(CH_2)_4-CH(NH_2)-COOH$ |
| Methionine | Met | M | $CH_3-S-(CH_2)_2-CH(NH_2)-COOH$ |
| Phenylalanine | Phe | F | $Ph-CH_2-CH(NH_2)-COOH$ |
| Proline | Pro | P | |
| Serine | Ser | S | $HO-CH_2-CH(NH_2)-COOH$ |
| Threonine | Thr | T | $CH_3-CH(OH)-CH(NH_2)-COOH$ |
| Tryptophan | Trp | W | |
| Tyrosine | Tyr | Y | $HO-p-Ph-CH_2-CH(NH_2)-COOH$ |
| Valine | Val | V | $CH_3-CH(CH_2)-CH(NH_2)-COOH$ |

## Table 6.7
### Classification and functions of some amino acid side chains

| Classification | Amino acid | Function |
|----------------|------------|----------|
| Hydrophobic/aliphatic | Leu, Ile, Val | Structural, space-filling, hydrophobic bonding |
| Hydrophobic/aromatic | Tyr, Phe, Trp | As above + pi bonding, ligand binding, UV absorbance |
| Charged/acidic/hydrophilic | Asp, Glu | Anion binding, H-bonding, electrostatic bonding |
| Charged/basic/hydrophilic | Lys, His, Arg | Cation binding, electrostatic interactions |
| Uncharged/hydrophilic | Glu, Ser | H-bonds, S–S bonding, active sites, post-translational modification |
| Amide, hydroxyl, sulfhydryl | Cys, Met | Disulfide (S–S) bond formation Enzyme active sites, binds heavy metal (Hg, Ag), antioxidant action |

The three-letter abbreviations for amino acids are given in Table 6.6.

Amino acid composition also affects **taste** and **flavor**. A number of amino acids have a distinctively **sweet** taste (Gly, Ala, Thr, Pro, Ser, and Glu) or **bitter** taste (Phy, Tyr, Arg, Leu, Val, Met, His). Two amino acids are associated with unami and **sour** taste (Glu, Asp). The taste of some fermented foods is related to their amino acid and peptide profiles. Cheese and miso (soy sauce) are just two examples of fermented foods whose flavor is believed to be largely caused by their amino acid composition.

A quantitative determination of amino acid composition involves counting the proportion of amino acid side chains that are aromatic, polar, nonpolar, acidic or basic (Table 6.7). Nonpolar amino acids have aliphatic carbon skeletons or aromatic side chains possessing benzene-type ring structure. Some R groups have one or more of the following chemical groups: hydroxyl (OH), sulfhydryl (SH), amino, and carboxyl. Some side-chain groups serve as sites for hydrogen bonding, electrostatic attraction or van der Waals' interactions between protein chains.

### 6.3.2.2 Post-Translational Modification and Primary Structure

**Translation is** the process of converting genetic information from mRNA into protein. Proteins intended for use outside of the cell (extra-cellular proteins) are produced on ribosomes attached to membranous tubes called the **endoplasmic recticulum** (ER). Newly synthesized proteins pass into the lumen of the ER and then proceed from there gradually towards the Golgi vesicle for packaging and export. It is during transit along the ER that proteins undergo a series of enzyme catalyzed reactions called post-translational modification (Table 6.8) leading to profound effects on their characteristics.

Post-translational modification of caseins involves **phosphorylation** and **glycosylation**. Phosphate or carbohydrate groups are attached to the OH group of Ser and Thr. For example the four **casein variants** from cow's milk, $\alpha_{S1}$, $\alpha_{S2}$, $\beta$, and $\kappa$-casein possess 8, 11, 5, and 1 phosphate group per molecule. One the most widely known **glycosylated** food proteins is $\kappa$-casein. Here the molecules are normally located on the external region of casein micelles. Steric interactions between the highly hydrated carbohydrate residues of $\kappa$-casein protect milk casein micelles from coagulation and ensure their colloidal stability. After treating $\kappa$-casein with the enzyme chymosin, the "carbohydrate" covering is removed allowing milk casein micelles to coagulate. Chymosin induced coagulation of $\kappa$-casein is crucial for cheese manufacture.

**Collagen** makes up a large proportion of the proteins found in connective tissue. Post-translational modification of collagen involves **hydroxylation**, which is the addition of new OH groups to the Pro residue. The reaction is catalyzed by **4-hydroxyproline hydroxylase**. The enzyme requires vitamin C for activity. Hydroxylation increases the hydrogen bonding capacity of collagen. The melting point for collagen increases with the degree of hydroxylation. A lack of vitamin C reduces 4-hydroxyproline hydroxylase activity, reduces collagen hydroxylation, and produces weakened connective tissue and many of the symptoms of **scurvy**, such as bleeding gums and weakened teeth. Post-translational modification of collagen

### Table 6.8
### Origins and types of covalent bonds in food proteins

| Covalent bonding | Comments |
| --- | --- |
| Peptide bonds | Amide bond formed between two amino acids during protein synthesis |
| Post-translation covalent bonds | Biochemical transformations within the endoplasmic |
| Disulfide bond formation | reticulum or Golgi apparatus generates |
| Glycosylation | multiple isoforms of proteins |
| Hydroxylation | |
| Phosphorylation | |
| Limited proteolysis | |
| Iodination | |
| Acetylation | |
| Organic chemical modification | Laboratory modifications applied to protein |
| Acylation | ingredients for technological aims |
| Glycation | |
| Phosphorylation | |
| Oxidation | |
| Proteolysis | |

is also catalyzed by the enzyme **lysine oxidase**. The enzyme catalyses the transformation of some Lys residues of collagen into aldehydes leading to a gradual formation of Lys–carbonyl cross-links. It may be expected that cross-link formation would result in increased toughness of long-lived connective tissue such as found in the skin. In summary, post-translational modification of collagen results in materials of a wide range of hydration characteristics and toughness. Hide collagen is a very tough material compared with very soft forms of connective tissue found in the eye lens or kidneys.

### 6.3.2.3 *In Vitro* Chemical Modification

Amino acids can be altered via purposeful chemical modification in a laboratory. *In vitro* chemical modification involves treatment with **acetic anhydride, succinic anhydride, phosphoric acid** or a mixture of acid/alcohols. Acyl, phosphate or methyl groups can be introduced by esterification, phosphorylation, and ether formation. Acetylation or succinylation transforms the side-chain amine group of Lys into acylated Lys and alters the electrical charge on this residue from +1 to −1. The charge on a protein molecule decreases by 2 units per every Lys group modified. Lys modification therefore alters the net charge and isoelectric point of a protein. Another simple form of chemical modification is to heat proteins with carbohydrates in the dry state leading to **glycated** proteins. Chemically modified protein ingredients have altered functionality compared to the unmodified proteins.

### 6.3.2.4 Process Induced Modification of Primary Structure

Process induced changes in primary structure occur when proteins are heated under slightly alkaline conditions. At first, there is a loss of HX (where X is SH, OH, $NH_2$) via **$\beta$-elimination**. An unsaturated C=C bond is then formed which undergoes electrophilic attack by an adjacent groups to form a cross-link. The reactions occur predominantly at Cys, Ser or Lys residues leading to so-called **lysinoalanine cross-links**. The eliminated groups can re-add to the C=C bond creating a mixture of geometric isomers in a process called **racemization**. Another important process induced reaction involves the –$NH_2$ group of lysine and the aldehyde group of carbohydrates. Proteins react with lactose during the **Maillard reaction** (Chapter 10). Process induced covalent bonds are usually unbreakable by enzymes in the digestive tract of non-ruminants.

### 6.3.3 SECONDARY STRUCTURE

Factors affecting protein conformation are discussed in Chapter 7 (Section 7.5). Small amino acids appear to be more compatible with $\alpha$-helix formation (helix formers). Very large side chains are easily accommodated within the beta-sheet structure. Pro is a **helix breaker**. Collagen with 30% Pro displays a left handed poly-proline helix. $\beta$-turns (alias reverse turns) are U-shaped loops that allow large proteins to fold into compact shapes. Here are some generalizations about $\beta$-turns: (1) up to 25% of the amino acids in a globular protein form $\beta$-turns; (2) $\beta$-turns contain the residues proline and glycine; (3) $\beta$-turns occur at the surfaces of proteins; and (4) $\beta$-turns contain four amino acids. Extended sequences of $\beta$-turns appear to be found in wheat proteins.

The secondary structure of some important food proteins are listed in Table 6.9. $\alpha$-helix content is inversely related to the quantity of $\beta$-sheet.[1] This is not surprising perhaps because wholly random proteins are rare. Moreover $\beta$-sheet and $\alpha$-helix cannot exist on the same segment of protein. Data from Table 6.9 fit the general function $\alpha$-helix (%) = $60.5 - 0.8213$ ($\beta$-sheet) with a correlation coefficient ($R$) value of 0.91. Myoglobin, serum albumin, and lysozyme contain the $\alpha$-helix structure whereas $\beta$-lactoglobulin, soy proteins, and

**Table 6.9**
**The range of secondary structure found in some proteins**

| Protein | % α-helix | % β-sheet | % β-turns | % Aperiodic | Other* |
|---|---|---|---|---|---|
| Myoglobin | 85.7 | 0.0 | 8.8 | 5.5 | 14.3 |
| Bovine serum albumin | 67.0 | 0.0 | 0.0 | 33.0 | 33.0 |
| Insulin dimer | 60.8 | 14.7 | 10.8 | 15.7 | 26.5 |
| Lysozyme A | 45.7 | 19.4 | 22.5 | 12.4 | 34.9 |
| Papain | 27.8 | 29.2 | 24.5 | 18.5 | 43.0 |
| α-Lactalbumin | 26.0 | 14.0 | | 60.0 | 60.0 |
| Bovine trypsin inhibitor | 25.9 | 44.8 | 8.8 | 20.5 | 29.3 |
| Ribonuclease A | 22.6 | 46.0 | 18.5 | 12.9 | 31.4 |
| Chymotrypsinogen | 11.0 | 49.4 | 21.2 | 18.4 | 39.6 |
| Phaseolin | 10.5 | 50.5 | 11.5 | 27.5 | 39.0 |
| Soy 11S protein | 8.5 | 64.5 | 0.0 | 27.0 | 27.0 |
| β-Lactoglobulin | 6.8 | 51.0 | 10.5 | 31.5 | 42.0 |
| Soy 7S protein | 6.0 | 62.5 | 2.0 | 29.5 | 31.5 |
| Immunoglobulin G | 2.5 | 67.2 | 17.8 | 12.5 | 30.3 |

*Techniques for determining % β-turns and aperiodic structure are not very precise, therefore they are lumped together as "other" secondary structure.

immunoglobulins are β-sheet proteins. Table 6.9 lists the range of secondary structures found in some proteins.

### 6.3.4 SUPER-SECONDARY STRUCTURE

Attractions between α-helices and/or β-sheets lead to super-secondary structure. This is the most sophisticated level of structure found within the fibrous proteins. Non-native super-secondary structures occur during food processing. The principal types of super-secondary structure are: (1) α–α (**coil–coil**) structure; (2) β–β (or **beta–meander**) structure; (3) α–β–α structure; and (4) β–α–β–α (**Rossmann fold**). Helices or β-sheets usually end in a reverse turn. For example, the coil–coil arrangement involves two **antiparallel** α-helices joined by a reverse turn. Specific secondary structure elements pack against each other as directed by primary structure repeat sequences.

Super-secondary structures are stabilized by **complementary interactions** between elements of secondary structure. To understand how this happens, we should note that helices are **nonsymmetrical structures**. For instance the α-helix has overall dipole moment along its long axis. The sides of an α-helix are also either hydrophobic or hydrophilic. β-sheets have nonpolar sides. The asymmetric nature of α-helices and/or β-sheets leads to interactions between complementary surfaces. Polar surfaces attract polar surfaces of opposite charge. Hydrophobic surfaces pack together to produce ordered arrays. Instructions for organizing super-secondary structure appear to reside in primary structure repeats. Recall that the collagen molecule has 1000 amino acids with the simple sequence Gly-X-Y$_{340}$. The myosin chain has a 7-amino acid and 15-amino acid repeat along is length. Indeed, most fibrous proteins exhibit some form of periodicity in their primary structure.

### 6.3.5 DOMAIN STRUCTURE

Domains are geometrically distinct regions within a protein. They appear to function as (1) evolutionary units, (2) functional units, and (3) folding units. In recent times, the Mendelian idea of "one gene, one phenotype" has undergone modification. Following the discovery of the DNA double helix, it was assumed that one gene encoded for one complete protein. The gene was represented by one uninterrupted piece of DNA. In 1977 it was discovered that eukaryotic genes consist of encoding sections of DNA (**exons**) interspersed with nonsense sequences called **introns**. The gene for myoglobin has two introns. The gene for chicken egg ovalbumin has seven introns whilst the collagen gene has 40 exons[a].

The **split gene** concept led to a notion that exons encode for functional units within proteins. For example, an exon that encodes for the heme binding domain may have been recombined with other exons to produce a host of heme proteins including myoglobin, cytochrome C oxidase, peroxidase, and chlorophyll). Other exon recombinations led to recurrent structural motifs such as enzyme co-factor binding sites and membrane binding sites. Exon recombination appears to be general mechanism that increases the rate of protein evolution as compared to point mutations.[2]

### 6.3.6 TERTIARY OR GLOBULAR STRUCTURE

Protein **tertiary structure is** the overall 3-dimensional shape adopted by a polypeptide chain. In general, the tertiary structure is synonymous with a globular state with a (1) spherical shape, (2) hydrophobic core, and (3) charged surface. Chain folding occurs such that all nonpolar amino acids are arranged on the inside of the globular "oil-drop" shape. Charged amino acid side chains are directed towards the outside of the globule where they form hydrogen bonds or weak electrostatic interactions with water. A more sophisticated model for globular proteins allows for nonpolar amino acids on the protein surface. The predominantly polar globular surface has hydrophobic patches. The extent of **surface hydrophobicity** shows a high correlation with certain functional characteristics. Protein surface hydrophobicity is routinely measured by fluorescence studies, and brief details are given in Appendix 6.2. The packing density of amino acid side chains in the globular protein has been calculated as 0.75–0.78 g/$cm^3$. The volume of a globular protein can be estimated from its molecular weight ($M$) using the formula, volume $= 1.3M$. The results for a 60,000 Da molecular weight protein are given in Table 6.10.

The density of a protein globule approaches that of solid crystalline material. That proteins also possess a fluid-like character is shown by the fairly rapid exchange between added deuterium molecules and hydrogen atoms from the protein interior. In other words, most proteins undergo rapid motions ranging from localized flip-flop motions of residues to wholesale spontaneous folding–unfolding motions. It is tempting to ascribe some of these characteristics to the presence of both crystalline and amorphous regions within proteins.

---

**Table 6.10**
**Formulae for estimating the dimensions of globular proteins**

| Protein parameter | Formula | Example |
|---|---|---|
| Volume ($V$) | $1.3\,M$ | $V = 78{,}000\,\text{Å}^3$ |
| Surface area ($A$) | $5.72\,M^{2/3}$ | $A = 1530\,\text{Å}^2$ |
| Surface:volume ratio (SV) | $4.44\,M^{1/3}$ | $SV = 0.113\,\text{Å}^{-1}$ |
| Diameter ($d$) | $1.35\,M^{1/3}$ | $D = 53\,\text{Å}$ |

M is molecular weight.

---

[a]Bacterial genes consist of linear pieces of DNA the size of which gives an accurate impression of the size of the corresponding protein. A bacterial protein consisting of 160 amino acids is likely to have a gene comprising 480 bases + 20 or so regulatory base sequences. We cannot perform similar calculations for eukaryotic proteins.

## 6.4  STABILIZATION OF PROTEIN STRUCTURE

### 6.4.1  ELECTROSTATIC INTERACTIONS

Electrostatic bonding involves attraction between acidic (aspartate, glutamate) and basic (lysine, arginine, histidine) amino acid side chains. Solvent characteristics (pH, ionic strength) affect electrostatic bonding by changing the ionization state of acidic and basic side chains. Salts, sugars, and other solutes alter the strength of charge–charge interactions by altering the **solvent dielectric constant**.

### 6.4.2  HYDROGEN BONDING

This type of electrostatic bonding occurs when hydrogen is shared between two highly electronegative atoms (N–H···O). The H-bonded groups carry a partial charge. The presence of high concentrations of Asn and Glu in cereal proteins allows high degrees of H-bonding. Hydrogen bonds increase at low temperatures and low polarity solvents. Internally placed hydrogen bonds may contribute to protein stability. Surface-placed H-bonds are normally disrupted by water, which is a strong H-bonding agent.

### 6.4.3  HYDROPHOBIC INTERACTIONS

Proteins fold in order to remove nonpolar amino acid chains from contact with water. This so-called hydrophobic effect is the most important type of stabilization for native proteins. The role of water in supporting hydrophobic interactions is paramount. Nonpolar amino acid chains are squeezed from contact with water thereby organizing a protein's shape. No actual interactions take place between individual nonpolar R-groups. Consider the thermodynamics of the hydrophobic interaction, that is, energy changes that occur as we transfer a nonpolar group into water (Table 6.4). Figure 6.4 explains the hydrophobic interaction.

Hydrophobic interactions account for the tendency of water and oil to **demix**. Figure 6.4 shows the transfer of a nonpolar amino acid side chain (R) into water. The equilibrium constant ($P$) and free energy change for transfer ($\Delta G_{tr}$) is also shown. For transfer to occur spontaneously, $\Delta G_{tr}$ should be negative. For the dissolution of methane in water, there is small heat evolved (negative $\Delta H$) due to transfer. The negative $\Delta H$ favors solubilization. The transfer of $CH_4$ into water is associated with a negative entropy change ($\Delta S$). The sum of the enthalpy and entropy changes leads to an overall *positive* Gibbs free energy change for transfer according the relation, $\Delta G_{tr} = \Delta H - T\Delta S$. The transfer of nonpolar groups *into* water does

Hydrophobic interactions

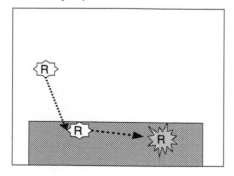

- $R + H_2O' \rightleftharpoons R.(H_2O)_n$
- $P = [R.H_2O]/[R]$
- $\Delta G_{tr} = -RT \ln P = \Delta H - T\Delta S$
- Negative $\Delta G_{tr}$ means transfer to water
- Observation: negative $\Delta H$ favors mixing
- Observation: large negative $\Delta S$ favors demixing

**Figure 6.4**  Explaining the hydrophobic interaction. Transfer of a nonpolar group (R) into water occurs by cavity formation and H-bonding to the solvent. $P$ = partition coefficient and ($\Delta G_{tr}$) = free energy change for transfer from oil to water.

not occur spontaneously because $T\Delta S$ is lower than $\Delta H$. In contrast, demixing $CH_4$ and water is a spontaneous process with a favorable (negative) $\Delta G$.

Several models have been proposed to explain the thermodynamics of the hydrophobic interaction. To transfer an R-group into water requires (1) creation of a cavity with sufficient volume to accommodate the dissolving nonpolar R-group; (2) development of interactions (probably H-bonds) between R and surrounding water molecules; and (3) distortion of interactions between water molecules. The large entropy decrease for solubilizing nonpolar R-groups, is explained by the formation of iceberg-like structures (**clathrates**) around each residue. Water molecules within a clathrate have reduced mobility compared to nonclathrate water. It is possible that water surrounding R-groups is compressed to make room for the R-group. Actually, thermodynamic considerations account for the sequestration of nonpolar groups away from water but our picture of processes leading to observed thermodynamic results are far from clear. The demixing process reduces the total cavity area needed accommodate R-groups within water. In support of this idea, experiments show that the strength of hydrophobic interactions is directly related to the size of amino acid side chains. We will encounter this idea later as a correlation between protein stability and the total accessible surface area removed from contact with water.

### 6.4.4 HYDROPHOBIC AMINO ACIDS

The **hydrophobicity** of amino acids was measured by Nozaki and Tanford as $\Delta G_{tr}$. When a solute is added a system containing two immiscible solvents (octanol and water are common examples) a certain fraction dissolves in both phases. The concentration of the test compound in octanol divided by the concentration in water ($C_O/C_W$) yields the **partition coefficient** $P$. For hydrophilic substances $P \ll 1$ whilst hydrophobic solutes yield $P \gg 1$. The following expression, $P = S_O/S_W$ is also worth noting. Here $S_O$ is solubility of the test compound in the octanol phase in the absence of water. We find $S_O$ by adding incremental amounts of solute to octanol until no more dissolves. In the same way, $S_W$ is the limiting solubility of the test compound in water. For convenience, values of $P$ are expressed on a logarithmic scale. The so-called "log $P$ scale" ranges from +4 for highly hydrophobic compounds ($P = 10,000$) to −4 for hydrophilic compounds ($P = 0.0001$). In another drive for convenience we can multiply log $P$ by the term $-RT$ to give $\Delta G_{tr}$:

$$\Delta G_{tr} = -RT \log P \tag{6.1}$$

where $R =$ gas constant and $T =$ temperature (K). At room temperature $RT$ is about 2477.5 (J/mole)[b]. For hydrophobic solutes $\Delta G_{tr}$ is negative.

### 6.4.5 PROTEIN HYDROPHOBICITY AND SURFACE HYDROPHOBICITY

Bigelow determined the average hydrophobicity for proteins in terms $\Delta G_{tr}$ averaged for all constituent amino acids:

$$H_\Phi = \frac{\sum_i^n \Delta G_{tr}}{n}. \tag{6.2}$$

---

[b]$\Delta G_{tr}$ was calculated as $-RT \ln P$ where $P$ is the ratio of solubilities of amino acid in an organic ($S_o$) phase to the solubility in water ($S_w$). Amino acids are relatively insoluble in organic solvents therefore $S_o$ was determined for water–ethanol mixtures and the values extrapolated to 100% alcohol.

**Table 6.11**

**Protein hydrophobicity in relation to surface functional properties**

| Protein | Surface hydrophobicity | Surface tension (dynes/cm) | Interfacial tension (dynes/cm) | Emulsifying activity m²/g | Bigelow number (cal/mole) |
|---|---|---|---|---|---|
| Bovine serum albumin | 1400 | 57 | 10.3 | 166 | 1230 |
| κ-Casein | 1300 | 54.1 | 9.5 | 185 | 1200 |
| β-Lactoglobulin | 750 | 59.8 | 11 | 151 | 1230 |
| Ovalbumin | 60 | 61.1 | 11.6 | 57 | 1110 |
| Conalbumin | 70 | 63.7 | 12.1 | 105 | 1080 |
| Lysozyme | 100 | 64 | 11.2 | 55 | 970 |

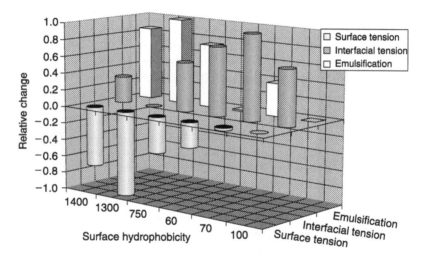

**Figure 6.5** Correlations between protein surface hydrophobicity and surface functional properties (surface tension, interfacial tension, and emulsification).

The Bigelow number ($H_\phi$) ranges from 870 cal/mole for tropomyosin to 1310 cal/mole for zein. $H_\phi$ is a measure of hydrophobicity for a completely structureless protein or for a protein hydrolysate. In contrast, the hydrophobicity for native proteins is affected by surface nonpolar groups. Kato and co-workers measured protein **surface hydrophobicity** ($S$) using fluorescence probes. Table 6.11 compares $H_\phi$ and $S$ values for several proteins and their relation to functional characteristics (Figure 6.5).

The following additional points help in further discussions: (1) hydrophobic interactions increase with temperature up to about 70°C; (2) hydrophobic interactions are directly correlated with the surface area of R-groups; (3) hydrophobic interactions are additive, with a net effect equal to the net surface area of different R-groups; (4) hydrophobic interactions are probably the most important interactions for protein structure; (5) hydrophobic interactions are also important for denatured protein structure (gels, foams, and emulsions); and (6) solutes (salts and sugars) alter solvent structure and hence hydrophobic interactions.

## 6.5  PROTEIN HYDRATION

Protein interactions with water affect their wettability, dispersibility, swelling, solubilization, viscoelasticity, gelation, emulsification, and foaming. Three classes of water are associated with proteins. **Chemically bound** water consists of isolated molecules that are bound to specific chemical groups through hydrogen bonding. **Structured or bound** water forms a monolayer around the protein molecule. On average proteins bind 30–50% water (by weight) following prolonged contact with an atmosphere with 90% relative humidity. Bound water is also described as unfreezable water on account of its lower freezing point compared with normal solvent water. The final class of water is bulk-hydration water. This is water that is swept along by the protein molecule as it moves. The hydration water has the same general characteristics as bulk solvent.

### 6.5.1 STAGES OF PROTEIN HYDRATION AND DISSOLVING

Protein–water binding can be displayed as an **adsorption isotherm**. This is an S-shaped graph showing the mass of water bound (y-axis) plotted against relative humidity of (RH, x-axis) of the atmosphere in contact with the protein. A value RH = 100% implies the presence of liquid water. Proteins adsorb about 0.3–0.5 g/g protein when brought into prolonged contact with air having a relative humidity of 90%. Next we need to consider protein interactions with liquid water. The ease with which protein powders can be dispersed in cold water is a function of their **wettability**. Fibrous and other semi-extended proteins (myosin, collagen, casein, seed storage proteins) form suspensions in cold water. The protein particles then swell before eventually going into solution. Unlike polysaccharides, raising the temperature does not solubilize protein powders owing to the tendency to aggregate.

### 6.5.2 FACTORS AFFECTING PROTEIN HYDRATION

Protein hydration is affected by intrinsic or protein-related factors, and environmental or extrinsic factors. Some of the major variables are listed in Table 6.12.

### 6.5.3 HOFMEISTER MODEL FOR SOLUTE EFFECTS

Salts increase protein solubility (salting-in) or decrease solubility (salting-out). The size of these effects follows the Hofmeister series, which shows how effectively serum albumin can be precipitated by different ions.[3,4] Table 6.13 implies that sodium, potassium or ammonium sulfate can be used as precipitation agents. By contrast, thiocyanate or chlorate salts would be more suitable as protein solubilizing agents.

### 6.5.4 PREFERENTIAL HYDRATION AND EXCLUSION

Some solutes stabilize proteins in solution. Other solutes destabilize proteins.[5] The effects can be explained by the effect of solutes on the surface tension of water. We assume that

---

**Table 6.12**
**Variables affecting protein hydration**

Intrinsic factors
- Protein structure—primary, secondary, super-secondary, etc.
- Net charge including effects of chemical and post-translational modification

Extrinsic or environmental factors
- Relative humidity, water activity, temperature
- Solution pH, ionic strength, salt type, presence of sugars
- Process history, e.g., heating

---

**Table 6.13**
**Hofmeister series of ions**

| Cations | | | $Ca^{2+}$ | $Mg^{2+}$ | | $Li^+$ | | $Cs^+$ | $Na^+$ | | $K^+$ | | $NH_4^+$ | |
|---|---|---|---|---|---|---|---|---|---|---|---|---|---|---|
| **Anion** | | $SCN^-$ | $ClO_4^-$ | | $I^-$ | | $NO_3^-$ | | $Br^-$ | $Cl^-$ | | $F^-$ | | $HPO_4^-$ | $SO_4^-$ |

| | Property | |
|---|---|---|
| | **Protein effects** | |
| ← Increasing | Solubility | Decreasing → |
| ← Increasing | Protein–water binding | Decreasing →* |
| ← Decreasing | Polymerization | Increasing →* |
| ← Increasing | Denaturation | Decreasing → |
| ← Decreasing | Stabilization | Increasing → |
| | **Solvent effects** | |
| ← Decreasing | Water structure | Increasing → |
| ← Increasing | Molar entropy for hydration | Decreasing → |
| ← Decreasing | Surface tension | Increasing → |
| ← Decreasing | Hydrophobic interactions | Increasing → |

*Effects observed at salt concentration sufficiently high to produce precipitation. Opposite effects account for increased hydration and stabilization.

denaturation leads to an increase in the surface area of a protein. The free energy for unfolding is described by Eq. 6.3:

$$\Delta G = \gamma \Delta A, \tag{6.3}$$

where $\gamma$ = surface or interfacial tension and $\Delta A$ is extra surface area needed to accommodate the unfolded protein. For protein stabilizing solutes, (1) there is increasing $\Delta G$ due to increasing values for $\gamma$, (2) there is anti-surfactant action leading to surface concentrations *below* those found in the bulk, and (3) the surface depletion of the added solute occurs due to **electrostatic exclusion** or **steric exclusion**. For instance, a solute with the same charge as a protein surface may be repulsed. **Steric exclusion** occurs if solute molecules are too large to fit easily within an array of water molecules at the protein–water interface, (4) anti-surfactancy therefore increases surface hydration resulting in preferential hydration, and (5) anti-surfactant solutes increase **hydrophobic** interactions **within** a protein. These considerations also predict that anti-surfactant molecules will reduce the wettability of suspended particles or increase particle transfer to a nonpolar phase.

The energy change necessary to introduce a nonpolar group into water ($\Delta G_{tr}$) is partly related to energy needed to create a cavity equivalent to the surface area, $\Delta A$. Arakawa and Timasheff suggested that stabilizing solutes are excluded from the immediate surroundings of a protein in a manner similar to the exclusion of solutes from the water–air interface. They termed this process **preferential hydration**. As the region surrounding a protein is depleted of solute molecules, there is a build-up of water molecules in the immediate surroundings of the protein compared to the bulk solution.

The precipitation and stabilization effects of ammonium sulfate on protein can seem contradictory. Protein precipitation is the result of heightened *protein–protein hydrophobic* interactions. The stabilizing action is the result of this additive's effect on *intra-protein* hydrophobic interactions (see above).

## 6.6 DENATURATION

### 6.6.1 DEFINITION AND SIGNIFICANCE

Protein denaturation was defined in the 1960s as the loss of secondary, tertiary or quaternary structure. This occurs when proteins are exposed to various stresses.[c] Denaturation is undesirable during the manufacture of protein ingredients. On the contrary, food manufacturing *requires* the destruction of the native protein state and restructuring to form gels, foams, and such esoteric structures like bread and cake. Loss of the native structure occurs during the manufacture of protein ingredients. Unit operations such as mixing, concentration, and drying, expose proteins to stresses like high temperature, shear, pH, ionic strength, and water content. Understanding and managing denaturation is important for the manufacture of optimal ingredients. Partial protein denaturation arising from preheating leads to desirable characteristics such as improved foaming. Processes such as emulsification, foaming, and gelation are preceded by a change in protein structure. Interestingly, the stability of protein is just sufficient to sustain the native structure under physiological conditions. Otherwise, proteins appear to be designed to undergo breakdown and resynthesis.

## 6.7 TECHNOLOGY OF PROTEIN INGREDIENTS

Mechanically pressed or solvent extracted **oilseed meals** are common sources of industrial proteins. Producers of industrial starch are also the largest producers of plant proteins, e.g., **wheat** or **corn gluten**. Food proteins also come from the **meat, fish,** and **poultry** industries in the form of high-protein waste; blood, offal, bone, skin, hoofs, fish trimmings, low value fish, and mechanically recovered protein. Protein ingredients are used to achieve a wide range of processing effects. Characteristics affected by food proteins are shown in Figure 6.6.

### 6.7.1 EXTRACTION OF PROTEINS FROM SOLID SOURCES

Protein sources can be extracted with organic solvent to remove fat. Using low temperature azeotropic distillation avoids protein denaturation, which leads to impaired solubility in water. **Proteolysis** is another way to extract proteins from suspensions. Digestion with enzymes reduces protein molecular weight and increases extractability. For example, partially hydrolyzed soybean protein isolate does not show isoelectric precipitation at a pH of 4–5. This material can be added to acidic fruit drinks for use as milk replacements. Endogenous or added proteases liquify proteins during traditional fermentations. Protein is also extractable **with alkali, acids or detergents**. Treatment of abattoir waste (lungs, rumen, intestines) with a solution of pH 10.5 solubilizes 50–80% protein. Alkali treatment can lead to significant losses of protein nutrient value due to the formation of the dehydroalanine and lysinoalanine racemization of amino acids, producing D-amino acids having no physiologic value. SDS and other detergents are alternative means for protein extraction. Removal of SDS from protein samples can be achieved by precipitation as the insoluble potassium salt; the protein sample is exposed to 40% ethanol plus 10% KCl.

### 6.7.2 RECOVERY OF PROTEINS FROM LIQUID SOURCES

High-protein effluents include cheese whey, blood, and waste water from potato processing. Blood for human food is collected under hygienic conditions. Dissolved proteins can be

---

[c]It is not certain whether different stress vectors can be inter-converted . It would be useful to know whether a 10°C rise in environment temperature equates to some specified pH change as regards the effect on proteins.

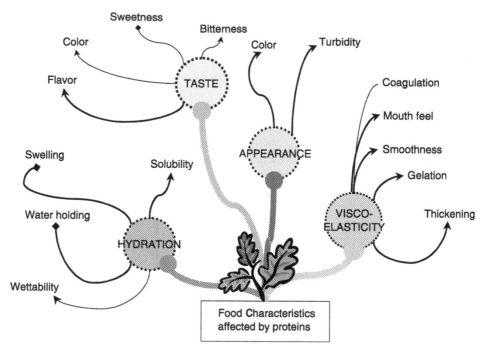

**Figure 6.6**   Characteristics affected by food proteins.

---

**Table 6.14**
**Forms of protein texturization**

| Mechanical methods | Others |
|---|---|
| • Thermoplastic extrusion | • Gelation |
| • Spinning | • Freeze texturization |
| • Steam texturization | • Film formation |
|  | • Foaming |
|  | • Emulsification |

---

recovered by **complexation** with pectin, alginic acid, and other **acidic polysaccharides** (lignosulfonic acid, carrageenan, carboxymethyl cellulose, and polyacrylic acid). Important variables are solution pH and ionic strength. **Low molecular weight coagulants** include $FeCl_3$, sodium hexametaphosphate, and sodium sulfate. Precipitation with alcohol and acetone is a widely used technique for isolating serum proteins. One advantage of **solvent precipitation** is the simultaneous removal of the heme group leaving a decolorized product. Blood proteins are also recovered by **heat coagulation** using **steam injection**. Heating induces protein denaturation and aggregation. Sometimes thermal treatment follows protein preconcentration by membrane ultrafiltration. Also worth mentioning is protein adsorption by **ion-exchange supports**. Nondenatured whey protein prepared via the ion-exchange method retains the characteristics of native proteins and is used as a replacement for relatively expensive egg white protein.

### 6.7.3   Texturization of Food Industrial Proteins

There are several processes for imparting texture to protein isolates. Table 6.14 shows some common methods of texturization. Extraction processes usually lead to protein products that need further processing to remove bulk water and increase its storage life. Drying, particle reduction, and texturization are some of the main unit operations for the manufacture of ingredients. These processes provide a convenient product for use in a range of food formulations.

### 6.7.4 Extrusion

A mixture of protein and water are forced through a heated barrel where it undergoes heating, compression, and mixing to form a thermoplastic mass. This is forced through a narrow die, followed by a sudden release of pressure that leads to rapid (flash) evaporation of steam. The resulting product has an expanded, puffed appearance. Undesirable flavors are lost yielding a bland product. Extrusion was successfully applied to range of food protein materials, especially cereal and soybean products. The technique is less suitable for milk due to the tendency for nonenzymic browning of lactose. Mixtures of cereals and milk powders have led to successful products.

### 6.7.5 Fiber Spinning

The technique developed by Boyer (1952) is based on textile technology. An alkaline protein extract (14–15% w/v; pH 10–11) is forced though a metal die having a large number of 0.1–0.2 mm holes. Fine threads of protein extruded into a bath of acidified brine (pH 3–4.5) where they coagulate. Protein fibers form which can be stretched and twisted to produce analogues for meat. Fats, polysaccharides, and flavors may added to modify product characteristics. Caseins are not very coagulable due to their low SH-group content, so they are co-processed with coagulable plant proteins such as soybean protein.

### 6.7.6 Gelation

Meat analogues were prepared from casein using a patented process developed by Anson and Pader (1957). Production of casein gels begins with an optional pre-heated stage that induces **casein** complexation with **whey proteins** and increases product yield by ~20%. Then acid curd is produced as usual. After washing to neutrality, the casein curd is coagulated by heating in presence of $CaCl_2$. Finally, the material is autoclaved leading to a chewy gel.

Casein gels are also produced by a modified acid or rennet coagulation method (see below). Fat filled milk is coagulated by rennet or acid. The curd is adjusted to neutral conditions and heated to high temperatures to produce a thermoelastic gel that can be molded and shaped for specific applications. Notice that this process has a great deal in common with normal cheese production.

The coagulation of casein using the enzyme **rennet** or by addition of acid leads to a weak curd for cheese. In the jargon of physical chemists, casein forms **particle gels**. These are described as enthalpy gels because the gelling subunits are discrete particles without much internal disorder. Shearing enthalpy gels does not impose a decrease in system entropy. Casein gelation is also the basis for **yogurt** formation. Here acid produced by lactic acid bacteria produces a weak casein gel. Adding polysaccharide thickeners can improve the characteristics of yogurt gels. The rate of acidification affects gel characteristics. Rapid addition of lactic acid to milk produces yogurt of inferior texture and a casein gel that is prone to syneresis. Production of chewy milk protein gels and of textured milk protein ingredients is shown in Figures 6.7 and 6.8, respectively.

Gelatin forms a **polymer gel** that is characterized by a 3-dimensional network involving the formation of junction zones by triple helices and some native cross-links. Interruption zones are formed where chain–chain separation occurs. The whole network entraps a great deal of water within its interstices. Polymer gels are thermo-reversible. Increasing the temperature

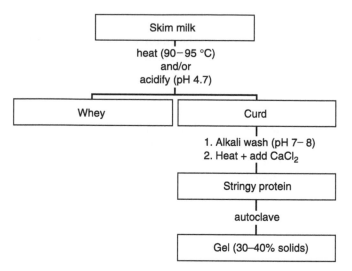

**Figure 6.7**  Production of chewy milk protein gels.

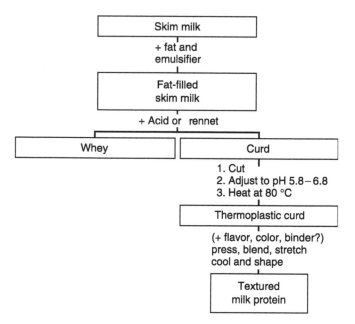

**Figure 6.8**  Production of textured milk protein ingredient.

leads to gel melting. Another important type is the **heat-set gel** formed by globular proteins. The 3-dimensional network now comprises partially deformable particles cross-linked by S–S bonds. Some junction zone formation is the result of intermolecular secondary structure. Globular protein gels are thermo-irreversible and are not disrupted at elevated temperatures.

### 6.7.7  PROTEIN FUNCTIONALITY

Functionality is any physical or chemical property that *affects* the characteristics of food during processing, preparation, storage, distribution, and consumption. We are concerned with hedonistic features like touch, feel, smoothness, roughness, and thickness. A list of functionality includes emulsification, foaming, film formation, gelation, and water holding

**Table 6.15**
**Variables affecting protein characteristics in foods**

| Factor | Comments |
|---|---|
| Intrinsic | Composition (presence of protein and nonprotein constituents), heterogeneity, protein structure and conformation |
| Manufacturing unit operations | Extraction, separation, concentration, drying, chemical modification, enzymatic modification |
| Conditions of use | Solvent characteristics, ionic strength, type of salts, temperature, pH, sugars, reducing compounds, lipids, starch |

(Figure 6.6). The food technologist is concerned with (1) identifying proteins with the requisite functionality, (2) cost of ingredients, (3) batch to batch variations in ingredients, (4) predictive tools for selecting, modifying, and/or designing protein ingredients, and (5) effect of formulation variables (presence of sugar, pH, salt) on product characteristics. Proteins are key ingredients in formulated foods.[6] Variables affecting protein characteristics in foods are given in Table 6.15.

## APPENDIX 6.1   THE OSBORNE EXTRACTION

To 1 g of wheat or rice flour add 20 ml of water and blend for about 30 s. Allow the mixture to stand for 5 min. Repeat the blending process 4–5 times and decant the water and keep. Next, add 20 ml of 70% ethanol to the residue flour and repeat the experiment. Decant this solution and follow with 20 ml dilute sodium hydroxide. This sequential extraction should recover albumin, globulins, prolamins, and glutelin. For optimum recovery of protein fractions each extraction can be extended to 90 min under gentle stirring with a magnetic stirrer.

## APPENDIX 6.2   MEASUREMENT OF PROTEIN SURFACE HYDROPHOBICITY

To a sample of protein dissolved in buffer, add a small amount of fluorescence compound, ANS (1-anilino-napthalene, 8-sulfonic acid) dissolved in water. Place the sample in fluorimeter, excite with light of wavelength 340 nm, and measure the fluorescence at 450 nm. The recorded fluorescence intensity is a measure of protein surface hydrophobicity. In a more robust experimental design, increasingly large amounts of protein are added to a fixed amount of ANS and the fluorescence is recorded. The slope of the plot of protein concentration versus fluorescence intensity is taken as a measure of surface hydrophobicity. Surface hydrophobicity correlates with functional properties, such as foaming and emulsification.

## References

1. Damodaran, S., Amino acids, peptides and proteins, in *Food Chemistry* (4th edition), Fennema, O., Ed., Marcel Dekker Inc., New York, 321–429, 1996.
2. Stryer, L., *Biochemistry* (4th edition), WH Freeman & Co., New York, 112–114, 1995.
3. Von Hippel, P. and Wong, K. Y., Neutral Salts: the generality of their effects on the stability of macromolecular conformations, *Science* 145, 577, 1964.
4. Hatefi, Y. and Hanstein, W. G., Solubilization of particulate proteins and nonelectrolytes by chartropic agents, *Proc. Natl. Acad. Sci.* 62, 1129, 1969.

5. Timasheff, S. N. and Arakawa, T., Stabilization of protein structure by solvents, in *Protein Structure: A Practical Approach*, Creighton, T. E., Ed., Oxford University Press, New York, 331–345, 1989.
6. Kinsella, J. N., Relationship between structure and functional properties of food proteins, in *Food Proteins*, Fox, P. F. and Condon, J. J., Eds, Applied Science Publishers, New York, 51–103, 1982.

# 7

# Principles of Food Material Science

## 7.1 INTRODUCTION

Material science principles are being applied increasingly to foods. The proportion of crystalline or amorphous structure within certain foods has important consequences on their mechanical behavior. This chapter introduces the principles of materials science applied to foods and related systems. Section 7.2 deals with free radical **chain reactions** for manufacturing packaging materials; similar reactions lead to the oxidation of lipids and other food components. (See also Chapter 9.) Consideration of polymer structure provides important insights about how food materials behave during processing and storage (Section 7.3). Section 7.4 deals with order–disorder transitions and their measurement in terms of the **glass transition temperature** ($T_g$) and the role of water as a **plasticizer**. Section 7.5 considers biopolymer structure and the connection between polymer shape, size, and the texture of liquid foods. Ideas in Section 7.6 are further expanded in Chapter 8, which describes food rheology.

### 7.1.1 CLASSIFICATION OF POLYMERS AND BIOPOLYMERS

A polymer is a large molecule made up of small repeating units called **monomers**. The term polymer, as introduced by Berzelius (1833), comes from the Greek *poly + mer* meaning *many* and *parts*. The size of a polymer is described by the degree of polymerization (DP). Molecules comprising 100 or more monomers (DP = 100) qualify as polymers. Two groups of polymers are distinguished by their response to heating: **thermoreversible** polymers melt during heating whereas **thermoset** polymers cannot be melted by heating. Polymers are also distinguished by the mechanism of polymerization. **Addition polymers** form via free radical **chain reactions** (Table 7.1) and **condensation** polymers form via **step reactions** (Figure 7.1).[1]

Chain reactions also occur with non free radicals species. For **ionic polymerization**, the reactive monomers are carbonium ions ($R_5C^+; R_3C^+$) or carbanium ions ($R_3C^-$). These reactions are used for the industrial production of rubber. The biosynthesis of **cholesterol** from isoprene ($CH_2=C(CH_3)=CH=CH_2$) units occurs via cationic polymerization. Another important chain reaction in food chemistry is the formation of thermoset gels by globular proteins under the influence of heat.

There are five recognized classes of synthetic polymers: fibers, plastics, rubber, adhesives, and coatings. These differ in their **stiffness modulus**: fiber > plastic > rubber. Another category of polymers is biopolymers derived from living cells including proteins (wool, silk), polysaccharides (cotton, cellulose), and plastics derived from modified cellulose. Digestible biopolymers (starch, plant, and animal proteins) are degraded by enzymes. Light weight and resistance to corrosion characterizes plastics. **Commodity plastics** are relatively cheap and intended for high volume use. Engineering plastics are low volume specialty materials.

---

**Table 7.1**

**Characteristics of polymerization reactions**

Free radical (chain) reactions, e.g., formation of vinyl polymers
- Polymerization via a number of growing chains
- DP very high
- Monomer depleted slowly, molecular weight increases quickly
- Initiation, propagation, and termination steps
- $R_p$ rises, remains constant, falls with initiation, propagation, and termination

Condensation (step) reactions, e.g., formation of polyesters
- Polymerization occurs throughout matrix
- DP is low
- Monomer is depleted quickly, molecular weight increases slowly
- One mechanism same throughout reaction

$R_p$ decreases steadily with time

DP = degree of polymerization. $R_p$ = rate of polymerization.

---

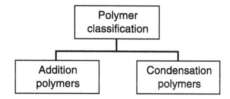

**Figure 7.1**   Classification of polymers.

## 7.2   KINETICS OF FREE RADICAL REACTIONS

The kinetics of free radical reactions is discussed in this section in relation to polymerization. The same principles apply to free radical mediated **oxidation of lipids**, vitamins, and other food components (Chapter 9). Here the focus is on chain polymerization by free radical processes or via condensation.

### 7.2.1   FREE RADICAL REACTION KINETICS

Free radical processes have three phases: initiation, propagation, and termination. Eq. 7.1 shows the formation of a free radical species starting from an initiator (*In*). **Initiation** involves energy input leading to the first appearance of highly reactive free radicals (*In\**) in the system. In Eq. 7.2, the activated *In\** species reacts with the monomer (*M*) to form an activated (free radical) monomer (*M\**).

$$\text{Initiation: } In \xrightarrow{k_i} In^*. \tag{7.1}$$

$$\text{Initiation: } In^* + M \xrightarrow{k_i} In + M^*. \tag{7.2}$$

It is generally assumed that the overall rate of polymerization ($R_p$) is determined by the rates of **initiation** ($R_i$) and **termination** ($R_T$). This idea has practical implications related to how we intervene in such reactions. (In Chapter 9 we consider free radical reactions for food oxidation leading to the development of off-flavors. Oxidation reactions can be prevented by decreasing the rate of initiation or increasing the rate of termination.) The rate of initiation is a function of the *initiator* concentration $[In]$ and the rate constant, $k_i$:

$$R_i = 2Fk_i[In], \tag{7.3}$$

where $F$ is a constant, the size of which is related to the **initiator efficiency**, which is a ratio of the number of radicals which start polymerization divided by the total number of radicals in the system. Equation 7.3 is based on a premise that the reaction from Eq. 7.1 is much *faster* than that shown in Eq. 7.2, which is the rate-limiting step for initiation. Furthermore the concentration of initiator $[In]$ is relatively constant and the kinetics of Eq. 7.2 is first order with respect to $[M]$.

It is time to consider the chain **termination** step, which is the reaction that ends free radical polymerization (Eq. 7.4). This is generally a reaction between two activated monomers ($M^*$). No distinction is made between $M^*$ and $M^*_{i+1}$ and consequently, the rate of termination ($R_T$) is expressed by a second order rate equation (Eq. 7.5):

$$\text{Termination: } M^*_{(i+i)} + M^* \longrightarrow M_{(i+2)}. \tag{7.4}$$

$$R_T = 2k_T[M^*]^2, \tag{7.5}$$

where $k_T$ is a second order rate constant. During the propagation phase (Eq. 7.6), the reaction enters a steady state:

$$\text{Propagation: } M^* + M_{(i)} \longrightarrow M^*_{(i+1)}. \tag{7.6a}$$

$$\text{Rate of propagation } (R_p) = k_p[M][M^*]. \tag{7.6b}$$

This means that the rate of formation and depletion of $M^*$ are equal during propagation: $R_i = R_T$ and consequently Eq. 7.7 applies. The concentration of activated monomer $[M^*]$ is then readily described by Eq. 7.8:

$$2Fk_i[In] = 2k_T[M^*]^2, \tag{7.7}$$

$$[M^*] = \left(\frac{Fk_i[In]}{k_T}\right)^{1/2}. \tag{7.8}$$

From Eq. 7.6b we can write the rate of propagation ($R_p$) as shown in Eq. 7.9 or 7.10:

$$R_p = k_p[M]\left(\frac{Fk_i[In]}{k_T}\right)^{1/2}; \tag{7.9}$$

$$R_p = k_p\phi[M][In]^{0.5}. \tag{7.10}$$

Eqs 7.9 and 7.10 predict that a graph of log ($R_p$) vs $[In]$ will yield a straight line with the gradient equal to 1/2. Heating the whey protein beta-lactoglobulin leads to a polymerization chain reaction via SH/S–S exchange. The initiator and monomer species are both molecules of beta-lactoglobulin and consequently $R_p \propto [M]^{1.5}$.

The rate of polymerization can diverge from predictions based on Eqs 7.9 and 7.10 for two reasons. In a highly viscous system, the rate of polymerization may exceed predictions based on these equations. Chain-end radicals are protected from termination but may continue to react with diffusible low molecular weight monomers. This so-called **Trommsdorff effect** or Norris–Smith effect is due to the reduced rate of termination of chain-end radicals due to their low mobility in a highly viscous system. The second form of deviation from predictions based on Eq. 7.9 is due to **chain transfer.** In this process, radical character is transferred from large to low molecular weight chains leading to a lower than expected DP and a broader range of polymer.

### 7.2.2 Types of Initiators

There are four classes of initiators for radical reactions: (1) peroxides or hydroperoxides, (2) azo compounds, (3) redox initiators, and (4) photoinitiators. **High temperature decomposition** of peroxides (ROOH), hydroperoxides (ROOR) or azo compounds ($R-N\equiv N-R$) generates free radicals. Initiation via redox mediators occurs due to 1-electron reactions. Eq. 7.11 shows the reduction of hydrogen peroxide by $Fe^{2+}$ (ferrous iron) to form the hydroxyl radical ($^\bullet OH$):

$$Fe^{2+} + H-O-O-H \rightarrow Fe^{3+} + {}^\bullet OH + OH. \tag{7.11}$$

A variety of transition metal ions ($Zn^{2+}$, $Cu^{1+}$, $Al^{3+}$) and peroxide/hydroperoxide species (e.g., cumyl hydroperoxides) form $^\bullet OH$. The mixture of peroxide and transition metal compounds is called the **Fenton reagent.** Ammonium persulfate is a common initiator for polyacrylamide gels:

$$^-O_3S-O-O-SO_3^- + X^+ \rightarrow SO_4^{\bullet} + SO_4^{2-} + X^{2+}. \tag{7.12}$$

**Photoinitiators** are light sensitive compounds. Examples include **disulfides** (Eq. 7.13) and **benzil** (Eq. 7.14). In both cases, illumination with ultraviolet or other radiation produces heterolysis and radical generation:

$$R-S-S-S + h\nu \longrightarrow 2RS^\bullet. \tag{7.13}$$
$$Ph.CO.COPh + h\nu \longrightarrow 2Ph.CO^\bullet. \tag{7.14}$$

### 7.2.3 Techniques for Free Radical Polymerization

Polymers can be prepared by (1) bulk polymerization, (2) solution polymerization, (3) suspension polymerization or (4) emulsion polymerization. With bulk polymerization, monomer and traces of initiators are processed together without solvent. Mass and heat transfer restrictions occur in the high viscosity mix leading to localized gradients in temperature and reactant concentrations. Solution polymerization, where reactants are dissolved in solution, improves mass and heat transfer. Reaction temperatures can be extended beyond the solvent boiling point by applying high pressure. Emulsion and suspension polymerization involve liquid or solid particles of monomer that are dispersed within a continuous solvent phase. Mixing and use of surfactants may be needed to achieve optimum results. Filtration or spray drying is used to recover polymer beads. Table 7.2 shows some polymers that are produced by chain-reaction polymerization.

### 7.2.4 Condensation Polymerization

Two monomers ($A + B$) react followed by the elimination of a small molecule, usually water. Polymer biosynthesis occurs at low temperatures in a relatively dilute solution within cells.

**Table 7.2**

**Some polymers produced by chain-reaction polymerization**

| Name | Monomer | Uses |
|------|---------|------|
| Polyethylene (PE) | $CH_2=CH_2$ (ethylene) | Bottles, tubing, packaging film |
| Polypropylene (PP) | $CH_2=CH.CH_3$ (propylene) | Household wares, artificial turf |
| Polyvinyl chloride (PVC) | $CH_2=CHCl$ (vinyl chloride) | Bottles, food wrap, piping, and hoses |
| Poly(tetrafluoroethylene) (PTFE) | $CF_2=CF_2$ (tetrafluoroethylene) | Nonstick surfaces, bearing, and insulation |
| Polystyrene | $CH_2=CH.Ar$ | Packaging, cups, ice buckets, foams, refrigerator doors |

**Table 7.3**

**Classes of polymers produced by condensation polymerization**

| Group | Monomer linkage |
|-------|-----------------|
| Polyether | M–O–M |
| Polysulfide | MSSM |
| Polyester | M–CO.O.M |
| Polyamide | M.CO.NH–M |
| Phenol, urea–formaldehyde resin | MOH, $MNH_2 + HCHO$ |

M, monomer.

It should be remembered that free radicals are an anathema to living cells where they cause damage to components like DNA, proteins, and lipids. Radical mediated processes appear to be involved in a range of disease states including cancer, arthritis, and eye cataract. There are sophisticated **antioxidant** mechanisms to control free radical by-products from cell metabolism. Within living cells polymer synthesis occurs almost exclusively via the condensation mechanism. To facilitate condensation polymerization, biosynthesis of polysaccharides, proteins, and lipids is coupled to the hydrolysis of the high-energy compound ATP. The kinetics of condensation polymerization conforms to a simple second order reaction. Classes of polymers produced by condensation polymerization are given in Table 7.3.

## 7.3 POLYMER STRUCTURE

### 7.3.1 INTRODUCTION AND SCOPE

The relationship between polymer structure and thermal mechanical properties is interesting for several reasons. Changes may occur in a material's stiffness as a function of temperature and addition of plasticizers. Dissolution, interactions with solvent, and polymer characteristics also affect solution viscosity. The behavior of polymeric systems is often determined by the following characteristics: (1) stereochemistry or tacticity, (2) backbone flexibility, and (3) intermolecular bonding.

### 7.3.2 POLYMER CONFORMATION AND TACTICITY

Polymers adopt a range of shapes or **conformations**, which can be reached by rotations at C–C bonds. Conformational changes also occur via the rearrangement of noncovalent bonds either within or between molecules. The energy needed for such transitions acts as a barrier between one conformation and the next. Elevated temperatures provide energy to overcome steric interactions between groups in physical contact. Note that **conformation** is different from chain **configuration**, which can only be changed by breaking C–C bonds.

Polymers sometimes contain asymmetric carbon atoms. According to their stereochemistry (see Appendix 7.1) polymers can be placed into one of three classes. **Isotactic** polymers have all side-chain groups in the same configuration. **Syndiotactic** polymers have alternating arrangement of side chains and **atactic** polymers have an irregular arrangement of side chains. **Tacticity** influences the type of packing between polymer chains, which in turn affects observed mechanical properties.

### 7.3.3 POLYMERS AS SOLIDS

The three states of matter (vapor, liquid, and solid) are distinguished by the spatial arrangement of molecules and the amount of vibrational **motion** between and within molecules. Molecules in a gas respond as a random ensemble. Molecules in a liquid or solid possess increasing levels of order. The crystalline state (defined below) exhibits perfect order at a temperature of absolute zero ($-273$ K). The degree of animation or **kinetic energy** possessed by materials varies in the order gas > liquids > solids[a]:

$$\text{vapor} \rightleftharpoons \text{liquid} \rightleftharpoons \text{solid}. \tag{7.15}$$

Cooling a gas leads to **condensation** and the formation of a liquid phase followed by freezing as the liquid is transformed into a solid. In contrast, increasing the temperature leads to **melting** followed by **vaporization**.

The term "phase" usually refers to the three major states of matter (Eq. 7.15). However, there are many gradations of molecular arrangements between the solid and the liquid phase. For example, a solid may contain two distinct phases arising from different geometric arrangements of molecules; carbon and graphite are classic examples of two different solid states for a single compound—polymorphism. Lipids also exist in different polymorphic states (Chapter 5). Some food materials are neither wholly liquid nor solids and are best described as soft solids, gels, pastes or glasses.

Solids exist in the **crystalline** or **amorphous** state. The molecules of a crystal are arranged into regular arrays. The neat arrangement allows strong interactions between neighboring molecules and the lowest degree of internal motion of all the solids types. The molecules of an amorphous solid have less spatial order, low attraction between molecules, and greater mobility as compared to those in a crystal. A high degree of **orientation** between polymers is synonymous with **high crystallinity**. Amorphous and crystalline regions may co-exist within a large polymer. Amorphous solids and liquids are both characterized as a disordered arrangement of molecules. Liquids formed by large linear molecules may possess significant order and are classed as **liquid crystals**.

### 7.3.4 FRINGE MICELLE AND LAMELLA MODELS

There are two models for polymer structure currently in circulation. The first is the **fringe micelle model**, according to which polymer materials contain regions of regular structure interspersed with disordered amorphous regions. The ordered regions contain areas of increased orientation between polymer chains. Strong **polymer–polymer** interactions lead to the formation of small crystalline phases or **crystallites**. These highly oriented regions extend over

---

[a]Recall that kinetic energy $= 0.5\,mV^2$, where $m =$ mass and $V =$ velocity.

relatively short lengths of polymer. More extended molecular arrays are formed by low molecular weight crystalline compounds. Within amorphous regions polymer–polymer interactions are low or nonexistent.

The alternative model for polymer materials is the **lamella model** which assumes that crystalline and amorphous regions occur within the single polymer chain. Crystallite regions are formed on a much smaller scale compared to the total length of a polymer. Apparently there is no simple way to distinguish between the fringe micelle and lamella models by experiment. The former applies to linear chain biopolymers, e.g., fibrous proteins (collagen/gelatin) and polysaccharides. The lamella model appears more likely for globular proteins (cf. discussions of super-secondary structure). Both models may apply in the case of starch.

## 7.4 ORDER–DISORDER TRANSITIONS WITHIN MATERIALS

At the **melting temperature** ($T_m$) materials absorb latent heat of fusion. Bonds between the ordered arrays of molecules are disrupted. The efficient packing within crystalline salts or sugars becomes disrupted and vacant spaces appear. These **vacancies** or imperfections lead to order–disorder transitions over a narrow temperature range. The melting process is all-or-nothing and is described as **cooperative**. By comparison, mixtures of differently sized macromolecules show less perfect crystal packing and the material melts over a broad range of temperature (cf. Chapter 5, Section 5.6.2 concerning fat plasticity and the effect of crystal structure on melting behavior).

From elementary chemistry, we know that $T_m$ is one indicator of the purity of chemical compounds. The presence of trace impurities reduces molecular order within a crystal. In consequence, the thermal energy needed to disrupt a crystal structure is reduced. The value of $T_m$ helps to confirm the identity of pure crystalline compounds; melting point apparatus is a common feature in most chemistry labs.

The opposite of melting is **crystallization**. Molecules from the liquid state become organized into an ordered array. The thermal bonding energy evolved during crystallization is called the **latent heat of crystallization**. Upon heating above the **glass transition temperature** ($T_g$) glassy solids are transformed into a rubbery amorphous state. The process, called **vitrification**, transforms a liquid to a glassy phase. Materials having both crystalline and amorphous regions produce successive order–disorder transitions with rising temperature. First, the glassy/amorphous/solid regions are transformed into a mobile rubbery phase. Then the crystalline regions melt[b].

### 7.4.1 TRANSITION TEMPERATURES

Large changes in material stiffness occur at the $T_g$ (glass transition) or $T_m$ (melting temperature). At temperatures higher than $T_m$ materials are transformed into a molten form, which can be shaped and fabricated into useful objects. Injection molding using the **screw extruder** or by compression molding is used to transform polymers into useful objects. The screw extruder has found widespread use for processing food materials into novel breakfast cereal products and a host of convenience foods.

Molecular motion is hindered in the glassy phase at temperatures below $T_g$. At temperatures below the $T_g$ we assume that an amorphous material exists as a glass. Within the glassy state, physical and chemical changes that require molecular diffusion become inhibited.

---

[b]Despite the glassy appearance of ice, frozen water is not a glass. In ice there is regular arrangement of water molecules. The melting of ice is a simple melting transition.

**Microbial growth** cannot proceed as nutrient and oxygen transfer to the growing cells is also limited. Furthermore, microbial cells entrapped in a matrix with amber-like consistency cannot divide. **Enzymatic reactions** and **chemical reactions** cease for similar reasons. Food products stored at temperatures below their $T_g$ are well (though not wholly) protected from deterioration via chemical, enzymatic, and physical process. Storage below $T_g$ enhances product **shelf-life**. At temperatures below the $T_g$, glasses are rigid and exhibit a tendency to fracture under stress. This characteristic leads to a **crispy texture** for low moisture food products such as cookies. Food dehydration is a relatively simple technology for raising the $T_g$ of most food materials above room temperature. The dependence of $T_g$ on moisture content of foods is described below. Characteristics that affect $T_g$ are listed in Table 7.4.

### 7.4.2   GLASS TRANSITION FOR AMORPHOUS SOLIDS

$T_g$ is defined as a temperature at which there is discontinuity in the temperature dependence of the partial molar volume, coefficient of thermal expansion *or* compressibility of a material. We can "unpack" this definition here and in Section 7.4.3 as follows. The **internal energy** for a material is made up of potential energy within covalent and noncovalent bonds and kinetic energy due to vibrations, bending, and other motions of atomic groups. Changes in internal energy are perceivable as changes in heat or else as nonthermal "work." The internal energy of any material can be *gauged* from its **specific heat capacity** ($C$; J/g/deg), which is the quantity of heat necessary to raise the temperature of a known mass of the substance by 1°C without a change in state. Where the material is pure water and the temperature change takes place from 14.5°C to 15.5°C, we call the resulting quantity the **calorie** (cal; J/g/°C). The specific heat (J/g/°C) for a range of solids increases with decreasing molecular weight. However, the average mole heat capacity is approximately constant for most solids $\sim 25$ J/mol/°C. For water $C_p = 75.24$ J/mol/°C. For 1 mole of material we refer to the mole heat capacity (J/mol/deg). The specific heat determined using an enclosed vessel (to inhibit volume expansion) is designated $C_v$. Using an open vessel we obtain the heat capacity at constant pressure ($C_p$). Because of the energy used for volume expansion, $C_p > C_v$ for any one material (Eq. 7.16):

$$C_p = \frac{TV\alpha^2}{\beta} + C_v \tag{7.16}$$

where $T$ is temperature (in kelvins), $V$ is the partial molar volume, $\alpha$ is the **coefficient of thermal expansion**, and $\beta$ is the **compressibility**.

---

**Table 7.4**

**Characteristics affecting $T_g$**

| Factor | Change in $T_g$ |
| --- | --- |
| Chain flexibility, C–C bond rotation | ↓ |
| Molecular weight | ↑ |
| Pendant groups and chain branching | ↓ |
| Polymer stereochemistry | ↑↓ |
| Hydrogen bonding and other noncovalent bonds | ↑ |
| Polymer blends (mixtures) and co-polymers | ↓↑ |
| Cross-linking | ↑ |
| Plasticizers | ↑↓* |

\* Raises $T_g$ compared to water as a plasticizer.

---

### 7.4.3 A PHYSICAL MODEL FOR THE GLASS TRANSITION

Compressibility is the *fractional* volume change experienced by a material when it is subjected to an infinitesimal change in pressure ($\partial P$). The thermal expansion coefficient is the corresponding volume change with a small increment in temperature ($\partial T$). The mathematical expression for these statements is

$$\beta = -\left(\frac{\partial V}{\partial P}\right)_T \frac{1}{V}, \qquad \alpha = \left(\frac{\partial V}{\partial T}\right)_P \frac{1}{V}, \tag{7.17}$$

where the molar volume ($V$) is the space occupied by one molecule. This quantity has three components: $V = V_{atoms} + V_{cavities} + V_{hydration}$. In Eq. 7.17, $\partial V/V$ is the fractional change in volume. You may notice that Eq. 7.17 describes the isothermal (constant temperature) compressibility and the isobaric (constant pressure) coefficient of thermal expansion coefficient. Allowing for temperature variations we measure **adiabatic compressibility** ($\beta^*$) from how materials interact with ultrasound when both pressure and temperature changes occur. Note that $\beta^* = \beta + \alpha \, (\partial T/\partial P)$.

Analysis of how $C_p$ changes with temperature provides a handle on the glass transition. For instance, differentiating Eq. 7.16 with respect to temperature leads to Eq. 7.18 which expresses dependence of $C_p$ on temperature. From Eq. 7.18 plotting a graph of $C_p$ vs $T$ should provide information about $V$, $\alpha$ or $\beta$ at different temperatures:

$$\left(\frac{dC_p}{dT}\right)_P \approx \frac{V\alpha^2}{\beta}. \tag{7.18}$$

A hypothetical $C_p$–$T$ graph is shown in Figure 7.2. The two straight lines in Figure 7.2 (A and B) are expected from Eq. 7.16 when the material under investigation is below or above the $T_g$. A break in the $C_p$–$T$ graph occurs if either $V$, $\alpha$ and/or $\beta$ changes suddenly. Recall our definition for $T_g$ as the temperature at which there is a change in $V$, $\alpha$ and/or $\beta$. The transition from glassy to an amorphous state (Section 7.4) occurs at the $T_g$. The glass transition will not produce a latent heat change because the average number of intermolecular contacts is the same in the glassy and the rubbery state.

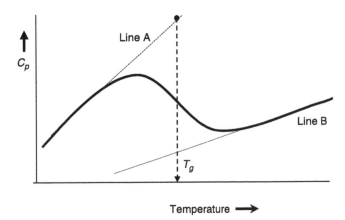

**Figure 7.2**  A schematic $C_p$–temperature trace from a differential scanning calorimeter.

Finally, we consider the effect of $T_g$ on the rheology of food polymers. A material's response to compressional stress is defined by its **bulk compliance** ($E'$) where

$$E' = \frac{\text{fractional volume change}}{\text{force/area}} = \frac{\text{strain}}{\text{stress}} = \frac{dV/V}{dP}. \tag{7.19}$$

Comparing Eq. 7.17 and Eq. 7.19 shows that the **bulk compliance** has the same units as compressibility. Bulk compliance is the inverse of elasticity measured as Young's modulus ($G$). $1/E = G$. Therefore we should expect changes in elasticity and other mechanical properties as a result of the glass transition. Food rheology is discussed further in Chapter 8.

### 7.4.4 TECHNIQUES FOR MEASURING $T_g$

Changes of $C_p$, $V$, $\alpha$ or bulk compliance occur at $T_g$. Rheological mechanical properties will also change as the temperature rises through the $T_g$ range. Techniques such as **differential scanning calorimetry** (DSC) and **dynamic mechanical analysis** (DMA) are convenient ways to measure $T_g$. During DMA analysis, we measure the complex modulus $G^*$ which has two components, $G'$ and $G''$. The magnitude of $G'$ reflects the elastic/solid character whilst $G''$ (the loss modulus) shows flow/liquid-like properties. The ratio of $G''/G'$ changes drastically at the glass transition temperature. These parameters are described further in Chapter 8.

### 7.4.5 CHARACTERISTICS AFFECTING $T_g$

Increasing **molecular weight** and cross-linking raises the $T_g$ but there is no effect of molecule size when DP > 200. Polymers that rotate freely at C–C bonds possess higher chain mobility and reduced $T_g$. **Pendant groups** or side chains, which hinder C–C rotation, increase $T_g$. However, polymer pendant groups can also reduce $T_g$ by hindering polymer chain–chain association whilst increasing penetration by solvent molecules. Polar groups increase $T_g$ owing to their tendency to increase **H-bonding** and ionic bonding. Indirect effects of polar groups lead to enhanced van der Waals' interactions arising from induced dipoles or permanent dipoles (electrostatic interactions). Polar groups will also improve affinity for water and thereby reduce $T_g$. Polymer tacticity increases $T_g$ according to the sequence, syntactic > idiotactic > atactic. With increased order of side-chain arrangements comes stronger polymer–polymer orientation and higher values for $T_g$. Clearly, it is not easy to predict changes in $T_g$ as a function of polymer characteristics (Table 7.4).

Mixtures of polymers (polyblends) and co-polymers have $T_g$ values that depend on the weight fraction ($w$) of each components and their **compatibility**. Mixing two compatible polymers yields a material of intermediate characteristics and one $T_g$ (Eqs 7.20 and 7.21). For two polymers that are incompatible, we obtain two distinct $T_g$ values:

$$T_g = w_1 T_{g,1} + k w_2 \frac{T_{g,2}}{w_1 + k w_2}, \tag{7.20}$$

$$T_g \approx \phi_1 T_{g,1} + \phi_2 T_{g,2}, \tag{7.21}$$

where $w_1$, $w_2$ = the weight fraction of polymer and plasticizer, and $k$ = constant. The terms $\phi_1$, $\phi_2$ represent the volume fractions of the two components.

### 7.4.6 EFFECT OF PLASTICIZERS ON $T_g$

Plasticizers cause polymers to be more plastic and malleable. For instance, polyvinylcholride (PVC) is normally a powder. To cast PVC into a film, one adds dioctylphthalate (DOP) as plasticizer. The plasticizer is distinct from a solvent (e.g., tetrahydrofuran) that may also be added though this evaporates from the PVC film as it sets. Platicizers have higher molecular weights (~300–1400 Da) compared to solvents. The principal effect of plasticizers on polymers is to reduce their $T_g$ in accordance with the **Gordon–Taylor equation** (Eq. 7.20). For biopolymers water is usually the plasticizer. Reducing the water content via dehydration increases the $T_g$ of food materials. The resultant decrease in molecular mobility is the basis for the longer shelf life of dehydrated food products.

## 7.5 BIOPOLYMER CONFORMATION AND SHAPE

The polysaccharides and proteins were described in Chapters 4 and 6. These biopolymers are subject to the same physical laws as the synthetic polymers. This section offers a rare opportunity to discuss proteins and polysaccharides together as well as comparing these with synthetic polymers. Most synthetic polymers contain one or two different monomers. Proteins contain 20 different amino acids. The monomer units for polysaccharides are 6-carbon and 5-carbon units of which there are 16 and 8 isomers, respectively. Patterns of covalent bonding between different monomer units leads to further complexity. The protein chain is built from the *peptide bond*. The glucose units of starch and cellulose are linked by $\alpha(1\rightarrow4)$ and $\beta(1\rightarrow4)$ *glycosidic bonds*, respectively. The difference in **linkage conformation** leads to profound differences between starch and cellulose. The former is storage polysaccharide and the means of storing solar energy within cells. Cellulose is nutritionally inert with a major role as plant cell wall material. Straw, wood, and cotton contain large amounts of cellulose. Some polymer science concepts are considered in relation to protein and polysaccharide structure below.

### 7.5.1 CONFORMATIONAL ANALYSIS

Any given polymer has large number of *potential* conformations. Consider a protein or polysaccharide with 100 monomers. If we assume each monomer can adopt two possible positions then the total number of potential conformation ($N$) is expressed by Eq. 7.22:

$$N = 2^{100} = 10^{30}. \tag{7.22}$$

Why do only a limited number of conformations appear in nature? The answer to this question is not fully known. However, the following generalizations can be made about biopolymer conformations: (1) folding occurs randomly in a manner that leads to the greatest number of favorable interactions between monomers, and (2) folding occurs in a way that minimizes the degree of internal strain within biopolymers.[2,3]

#### 7.5.1.1 Principles of Conformational Analysis

The principles of biopolymer folding were deduced by building physical and virtual models using computers. Bonds between monomer units were then rotated. For each hypothetical structure generated the total **conformational energy** ($E_C$) is determined as described below. The probability that a specific structure will exist in nature ($P_C$) was then determined by Eq. 7.23:

$$P_C = \exp(E_C/RT). \tag{7.23}$$

Therefore when $E_C \approx 0$ then $P_C \approx 1$. By contrast, when $E_C$ is large $P_C$ becomes progressively smaller and the corresponding conformation is less likely to appear.

There are four (de)stabilizing contributions to the total energy for any model conformation[c]. The net stability of any conformation ($E_C$) is dependent on van der Waals' interactions ($E_v$), electrostatic interactions ($E_e$), hydrogen bonding ($E_h$), and torsional energy ($E_\tau$):

$$E_C = E_v + E_e + E_h + E_\tau. \tag{7.24}$$

**Stabilizing** interactions are assigned a negative value whilst **destabilizing** interaction energy is positive. The chemistry of these weak interactions is well understood. The total interaction energy between atoms in a polymer chain can be readily calculated using the following equations from physical chemistry:

$$E_e = e^2 Q_1 Q_2 / Er, \tag{7.25a}$$

$$E_h = 49.2/r^6 + 2.3 \exp{(-3.6r)}, \tag{7.25b}$$

$$E_\tau = \sum 0.165(1 + \cos 3\theta), \tag{7.25c}$$

$$E_v = \frac{B}{r^{12}} - \frac{A}{r^6}. \tag{7.25d}$$

The **Leonard–Jones function** (Eq. 7.25d, cf. Eq. 8.11) shows the size of van der Waals' interaction energy ($E_v$) possible between two atoms separated by a distances $r$. Torsional energy (Eq. 7.25c) takes account of interactions between side chains during rotations at C–C single bonds. For instance as the C–C bond of ethane ($CH_3$–$CH_3$) rotates the total energy changes as a function of the torsional angle ($\theta$). The **eclipsed** and **staggered** conformations have high and low energy respectively. The staggered conformation is more stable as this leads to a wider separation of pendant hydrogen atoms on the adjacent carbon units (Figure 7.3).

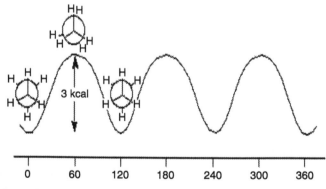

**Figure 7.3** Effect of C–C bond rotation on the total energy (y-axis) of ethane. Torsional energy is due to the staggered/eclipsed arrangement of pendant hydrogen on C1 and C2. The low or high-energy conformation of ethane occurs when C–H bonds are eclipsed and staggered arrangements, respectively.

---

[c]Recall that transitions between different conformations occur without disrupting covalent bonds, therefore contributions from these are ignored in this discussion.

### 7.5.1.2  Results of Conformational Analysis

The results from conformational analysis are conveniently presented as energy contour maps. The most well known of these is the **Ramachandran plot** for proteins. A conformation energy map has two axes labeled $\phi$ (phi) and $\Psi$ (psi). Points are drawn on the map for values of $\phi$ and $\Psi$ for which there is equal conformational energy. Contour lines are then drawn through points of equal energy value. The energy contour map for the disaccharide maltose was studied by **Rees and Smith**. Their results for maltose showed that:

(1) Only 3–5% of available values $\phi$, $\Psi$ resulted in high probablility structures.
(2) van der Waals' interactions measured as $E_V$ made the most significant contribution to $E$ and $P_C$. This result was achieved by simply deleting each of the four terms (one at a time) from Eq. 7.24 and observing the effects on the final calculations. In Eq. 7.24 most terms ($E_e + E_h + E_\tau$) could be deleted without great effect on the value for $E_C$. van der Waals' interactions are a direct measure of **steric interactions** between atoms. The results for maltose were assumed to apply to polysaccharides.
(3) Calculated values for $\phi$ and $\Psi$ agreed with experimental results from X-ray fiber diffraction studies using solid maltose.[2,3]

The preceding analysis implies that steric crowding around the glycosidic bond **restricts** chain folding and eliminates many *potential* conformations. Superimposed on this will be solvent effects. In solution, observed values for $\phi$ and $\Psi$ differ from those determined by modeling which does not take account of solvent interactions with the polymer. This is a major shortcoming of modeling especially as we consider the conformation of larger polymers.

As mentioned above, energy contour maps for proteins are called **Ramachandran** plots in honor of Professor G. N. Ramachandran who pioneered conformational analysis in the 1960s. Model dipeptides formed from pairs of 20 different amino acids showed the following recurring themes:

(1) Ramachandran plots for dipeptides have large regions of free space. As seen with polysaccharides, only a few bond angles measured in terms of values for $\phi$ and $\Psi$ produce low energy, high probability, structures.
(2) Amino acids having bulky side-chain groups experience increased folding restrictions and a limited range of useful $\phi$ and $\Psi$ values. Certain narrow ranges of dihedral angles $\phi$ and $\Psi$ are associated with the formation of secondary structures such as $\alpha$-helix, $\beta$-strands, and the left-handed poly-proline helix.
(3) Different amino acids have a propensity for particular secondary structures. This information has been useful in attempts to predict protein structure from sequence information.

### 7.5.2  PRIMARY STRUCTURE

The general view is that the primary structure contains all the information needed to determine higher orders of polymer structure. The primary structure for proteins was described in Chapter 6 (Section 6.3.2). At first, there seems little merit in discussion of the primary structure of polysaccharides. Amylose, amylopectin, glycogen, and cellulose all have "the same" sequence of glucose units but are very different materials. Apparently, the sequence of subunits is not a sufficient description for the primary structure for polysaccharides. According to Aspinall, polysaccharide primary structure encompasses seven elements (Table 7.5).

To determine the primary structure of a polysaccharide, we start with a disaccharide which is then is oxidized or reduced. The resulting disaccharide alcohol or acid is then hydrolyzed

**Table 7.5**
**Polysaccharide primary structure**

Identity of different sugars (monomers) present
Ring sizes (either pyranose or furanose rings)
Anomeric configuration ($\alpha$-glucose versus $\beta$-glucose)
Glycosidic bond linkage type (1→2, 1→3, 1→4)
Sequence of sugars
Substituent groups (e.g., sulfate, phosphate) and substitution points
Absolute configuration (D or L sugars)

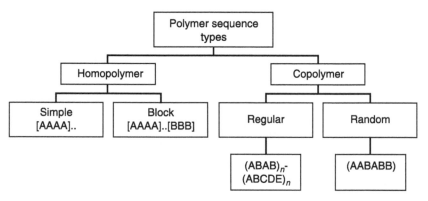

**Figure 7.4**  Biopolymer sequence types.

and the unmodified sugar is identified. In this way the sequence of sugars can be pieced together slowly. Simple sugars can then be identified via ion-exchange chromatography using an Aminnex® column coupled with a refractive index detector. Sugars or amino acids can also be transformed into relatively volatile derivatives and then analyzed by using gas chromatography–mass spectrometry (GC–MS). Techniques for assessing polysaccharide structure are summarized by McNeil et al.[4] and also by Aspinal.[5]

### 7.5.2.1  Types of Primary Structure

There are two classes of biopolymer sequence (Figure 7.4). **Homopolymers** contain one type of monomer. Examples of homopolymers include **starch** and **cellulose**. This set also includes polymers having two or more different monomers but where these are clearly separated along the polymer chain. **Alginate** and **pectin** are examples of block homopolymers. The second type of primary structure is the **copolymer**. As described elsewhere, proteins like collagen and cereal storage proteins have a simple repeating primary structure similar to regular copolymers.

### 7.5.2.2  Primary Structure and Biopolymer Function

A simple repeating primary structure allows the formation of regular higher order, structures such as fibers, ropes, and films. The fibrous proteins (tendon collagen, wool keratin, and silk fibrin) possess simple repetitive primary structures as does cellulose. In the unmodified state homopolymers are tough and insoluble in cold water. These characteristics make them suited for their biological role as structure forming elements in living organisms, e.g., wood, muscle, skin covering, nails, and hoofs. These materials have also been exploited as building materials (wood, straw) and clothing (plant and animal fibers). Improved understanding of biopolymers is beginning to impact on the design and manufacture of novel foods such as confectionery, desserts, and meat products, and analogues.

### 7.5.3 SECONDARY STRUCTURE

The local pattern of folding depends on the relative orientation of adjacent monomers determined by short-range (specific) interactions. Intrachain hydrogen bonds stabilize the $\alpha$-helix of proteins. Protein $\beta$-pleated sheets are stabilized by interchain hydrogen bonds. Polysaccharide secondary structure is determined by the relative orientation of two sugars joined by a glycosidic bond. Different secondary structures arise from different **linkage conformations**, which are sets of values for the dihedral[d] angles, $\phi$, $\Psi$, and $\omega$. As the angles of rotation are repeated from one residue to the next, biopolymers adopt a helical fold. All forms of secondary structures are helices described by three helix parameters: $n$ (the number of residues per turn), $d$ (vertical distance between adjacent monomers), and $p$ (pitch $= n \times d$). Factors that determine biopolymer secondary structures are presented in Section 7.5.1.

#### 7.5.3.1 Protein Secondary Structure

The most frequent protein secondary structures are the $\alpha$-helix and $\beta$-sheet. Different secondary structures are identified from their **helix parameters** (Table 7.6 and Figure 7.5).

Even a short polypeptide has a very large number of theoretical conformations (Section 7.5.1). Nevertheless, a unique native conformation with minimum free energy forms spontaneously. The rate of folding is much faster than expected assuming that polypeptide chains undergo random searches throughout all theoretical shapes. Whatever the pathway for folding, the native secondary structure is determined by very short range interactions. Specific types of secondary structures form which minimize unfavorable steric interactions within the polypeptide chain and optimize noncovalent interactions between side chains.

#### 7.5.3.2 Polysaccharide Secondary Structure

According to Rees and co-workers (1985) polysaccharides adopt four types of secondary structure in the solid state. Each kind of polysaccharide secondary structure has a distinctive set of helix parameters (Table 7.7). The **type-A helix** is a ribbon structure formed by structural polysaccharides including cellulose, hemi-cellulose (polyxylanose) or pectin. Alginate and carrageenan also adopt a type-A structure. This is formed by polysaccharides having the $\beta(1 \rightarrow 4)$ linkage. The family of $\beta(1 \rightarrow 4)$ glucan has an extended chain conformation that allows efficient strong chain–chain packing. The $\beta(1 \rightarrow 4)$ glucans exhibit relatively strong inter-chain hydrogen bonding and exclusion of water. Polysaccharide chain segments

---

**Table 7.6**
**Helix parameters for protein secondary structure**

| Type | $n$ | $d$ | Size (number of amino acids) |
|---|---|---|---|
| $\alpha$-helix | 3.6 | 1.5 | 10–15 |
| $\beta$-sheet | 2 | 3.47 | 3–10 |
| Polyproline | 3.12 | 3.12 | 1000 |
| B-turn | 3 | – | – |

---

[d] A dihedral angle is an angle formed by two planes meeting at an edge. As applied to biopolymers we suppose that the monomer units (sugars or amino acids) comprise atoms that all lie within a rigid plane, somewhat like a playing card. Chain conformation arises as these planes are rotated with respect to each other along their edges.

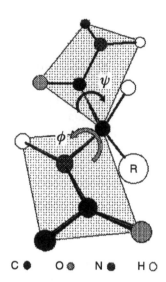

C ●    O ◉    N ●    H ○

**Figure 7.5** A diagrammatic presentation of a polypeptide segment showing points of rotation. The shaded plane rotates as one unit. Steric interactions occur between amino acid R-groups (triangles), C=O, and N–H groups.

**Table 7.7**

**Forms of polysaccharide tertiary structure in the solid state**

| Structure/examples | Properties |
|---|---|
| **Type A. Extended ribbon structure** | Fibrous, structural polysaccharides |
| Cellulose, hemi-cellulose pectin, alginate | Insoluble and tough |
| • General structure | $\beta(1 \rightarrow 4)$ glucan |
| • Helix parameters | $n = 2$ residues per turn, $d = 5.15$ Å per residue, and pitch $= 10.3$ Å |
| • Glycosidic bond angle | Obtuse C1–O–C4 angle ($> 90°$) |
| • Linkage conformation | $\phi = 210°$, $\Psi = 150°$ |
| **Type B. Hollow helix** | Storage polysaccharides, compact, relatively water soluble, easily mobilized for energy |
| • Starch, glycogen | |
| • General structure | $\alpha(1 \rightarrow 4)$ glucan |
| • Helix parameters | $n = 7$–8 residues per turn, $d = 0$ Å |
| • Glycosidic bond angle | Acute C1–O–C4 angle ($< 90°$) |
| • Linkage conformation | $\phi = 120°$, $\Psi = 130°$ |

possessing the extended helix conformation have low water solubility. The presence of type-A helical segments leads to high rigidity and characteristic ratio, root mean square end-to-end distance, and the intrinsic viscosity as discussed in Section 7.6. The presence of non-type A sequences, also present in the structure of pectin, alginate, and xanthan, account for their higher solubility compared to cellulose.

The **type-B structure** is formed by storage polysaccharides (amylose, amylopectin, and glycogen). The type-B helix is relatively compact with a large number ($n = 8$) of residues per turn. The hollow helix is relatively water soluble and unstable in solution unless present as a double helix. The type-B helix designation should not be confused with the description of cereal and tuber type-A or B starch. Another important amylose helix is $\gamma$-amylose (Chapter 4, Section 4.4.2). The **type-C helix** or flexible coil is formed by an expanse of monomers joined by

$\alpha(1 \rightarrow 2)$ linkages. This structure is expected to show substantial steric hindrance and a low probability of occurrence. The previous structure types should be considered examples of limiting behavior. A vast range of polysaccharide function is possible depending on the detailed chain formation.

## 7.6  SOLUTION TERTIARY STRUCTURE

The shape of a biopolymer in solution, gel or solid states is called **tertiary** structure (Table 7.8). Dry biopolymers have a **microcrystalline** and **amorphous** tertiary structure (Section 7.3.3). In the dissolved state, polymer–polymer contacts are replaced by polymer–solvent interactions. Polysaccharides and structural proteins adopt a **random conformation** in the dissolved state. Heat may be needed to promote dissolution. Starch dissolves readily with hot water or cold dimethyl sulfoxide. Proteins show more diverse dissolution behavior as described in detail elsewhere in this book.

The tertiary structure adopted by **native** globular proteins is spherical and retained after they crystallize from solution. Globular proteins show strong intra-molecular (within-polymer) interactions combined with an ability to retain a surface layer of adhering water. The importance of this hydration water is evident since dehydration (by spray drying or freeze drying) disrupts surface water, hydrophobic interactions leading to denaturation. Native fibrous proteins like collagen, are believed to exist in coil–coil form of super-secondary structure. Because the coil–coil motif extends over a large number of amino acids, they are sometimes classed as tertiary structure. It may be supposed also that the tertiary structure of "native" polysaccharides have yet to be determined for solids. In the next section we consider polysaccharide structure in solution (Table 7.8). These models also apply to globular proteins dissolved in 8 M urea.

### 7.6.1  RANDOM WALK MODEL

The main features of the random walk model are (1) polymers are composed of $N$ monomers each with an effective size ($L$), (2) **bonds** between monomers are **freely rotating,** and (3) the end-to-end dimensions of a random polymer ($R$) can be described in relation to the root mean area squared ($R^2$) traversed by a particle undergoing random Brownian motion. After $N$ random "jumps" *each* one covering a distance $L$, it can be shown that[6]

$$R^2 = NL^2, \tag{7.26}$$
$$R = LN^{1/2}. \tag{7.27}$$

The magnitude of $R$ estimated from Eq. 7.27 is usually lower than the experimental end-to-end distance ($R_{ex}$) for three reasons: (1) bonds between monomers are usually not freely rotating, (2) two atoms cannot co-exist in the same space due to so-called **excluded volume effects**, and (3) the polymer binds to solvent and thereby expands in size. The ratio of

---

**Table 7.8**
**Polysaccharide tertiary structure in the solid, sol, and gel states**

| Phase | Solid powder suspension $\xrightarrow{heat}$ | Sol | $\xrightarrow{cool}$ | Gel 3-dimensional network |
|---|---|---|---|---|
| *Tertiary structure* | *Crystalline amorphous (native structure)* | *Random coil, rigid rod-like structure* | | *Helical junctions + amorphous/disordered regions* |

**Table 7.9**
**The characteristic ratios for some polymers**

| Polymer | $C_\infty$ |
| --- | --- |
| Freely rotating polymer | 2 |
| Poly(ethylene oxide) | 3.4 |
| Cellulose | $\sim$100 |
| Cellulose ether | 8 |
| Polystyrene | 10 |
| Starch | 8–10 |

experimental and observed polymer sizes is called the **characteristic ratio** ($C_\infty$). The characteristic ratio increases due to steric restrictions, and excluded volume and solvent effects.

$$C_\infty = \frac{\langle R_{\text{ex}} \rangle^2}{\langle R \rangle^2}. \tag{7.28}$$

$C_\infty$ increases with polymer rigidity. Table 7.9 shows the characteristic ratio for a range of synthetic and natural polymers.

### 7.6.2 The Equivalent Sphere Model

A polymer and any bound solvent are modeled by an equivalent sphere with radius of gyration ($R_g$). For a completely rigid rod-like polymer, $R_g$ is the radius of gyration about the center of gravity raised to the second power. For a flexible polymer described by the random walk model, the radius of gyration is proportional to the experimentally measured end-to-end size ($R_{\text{ex}}^2$) where

$$R_{\text{ex}}^2 = 6R_g^2 \tag{7.29}$$

From the previous definition of characteristic ratio,

$$C_\infty = \frac{\langle 6R_g \rangle^2}{\langle R \rangle^2}. \tag{7.30}$$

### 7.6.3 The Worm-Chain-Like Model (Reptation Model)

A concentrated polymer solution can be compared with a plate of worms. Each polymer wiggles through a transient sheath, formed by its neighbors, via a worm-like motion. The self-diffusion coefficient of polymers is inversely related to the square of the molecular size ($D \propto 1/M^2$). The bulk modulus of a polymer melt is directly related to $M^2$ ($K \propto M^2$).

## 7.7 SOLID STATE TERTIARY STRUCTURE

The tertiary structure of most polysaccharides has yet to be characterized in the solid state. X-ray diffraction results for starch show characteristic patterns indicative of type-A and type-B starch. This information is apparently related to relatively short-range ordered (crystalline) structure. The special disposition of biopolymers within the starch granule and other polysaccharide solids remains an area of contention. Models of polysaccharide solid state structure need to contend with the arrangement of these biopolymers as supermolecular complexes within cell organelles, e.g., starch granule and plant cell wall. Protein structure in the native state is discussed in Chapter 6.

## 7.8  TECHNIQUES IN FOOD MATERIAL SCIENCE

As described in the previous section, polymer solution chemistry provides an important framework for analyzing food biopolymers. Hydrodynamic size can be studied by viscometry, ultracentrifugation or light scattering. Concentrated solutions, gels or so-called soft solids can be characterized via rheological studies involving destructive testing (Instron Universal testing) or small deformation (non-destructive) testing also called oscillatory rheology (Chapter 8). Studies of viscoelastic behavior provide structural information within specific frequency domains related to molecular dynamics, polymer–polymer, and polymer–solvent interactions.

The following techniques are commonly applied for the study of food materials:

(1) **Differential scanning calorimetry**. The specific denaturation enthalpy (cal/g) and onset temperature is related to residual ordered structure providing a basis for quality control and/or new structure–functionality relations.

(2) **Fourier transform infrared analysis**. This is an affordable method for estimating the degree of crystalline and amorphous structure within biopolymers in solution, gel or solid states.

(3) **Multiangle light scattering**. This technique used in conjunction with SEC (size exclusion chromatography) provides a weight average molecular weight distribution and root mean square size of hydrated polymers.

(4) **Fluorescence analysis** of surface hydrophobicity.

(5) A range of *wet chemical methods* for analyzing chemical residues.

## APPENDIX 7.1 STEREOCHEMISTRY

Stereochemistry (*ster'eokemistri*) is a branch of chemistry concerned with differences in the relative positions in space of atoms in a molecule and its effect on their physical and chemical properties. A key notion in stereochemistry is the asymmetric carbon atom (C*), which is any carbon atom that is bonded to four different groups or atoms. Compounds having one asymmetric atom have two non-superimposable structures: I and its mirror image I'. A molecule with two asymmetric atoms has 4 $(=2^2)$ possible stereochemical structures. Large folded molecules (e.g., coils or helixes) tend to have no clear line of symmetry and consequently I and I' will be non-superimposable (see Chapter 1 for further discussions).

## APPENDIX 7.2 CHANGES OF STATE

Heat supplied to any material either changes its state or changes its temperature. A change of state (Eq. 7.15) requires the absorption of **latent heat of vaporization** (fusion and/or sublimation). The latent heat ($\Delta H_L$) is defined as the quantity of heat necessary to induce a change of state for a unit mass of material from a liquid to gas (or from liquid to solid, or from solid to gas) without a change in temperature. $\Delta H_L$ is the energy that is utilized for breaking or forming bonds between molecules. Thus, $\Delta H_L = \frac{1}{2}nNE$, where $N$ = numbers of molecules per unit mass ($g^{-1}$), $E$ = bond energy (J) and $n$ = number of neighboring contacts per molecule. Changes in temperature involve energy that is utilized for altering the kinetic energy of molecules that may or may not cause a phase change.

There are two opportunities for confusion. First, melting can be easily confused with the dissolution by a solvent. Eq. 7.15 describes phase transitions for pure compounds in the absence of solvent. Second, vaporization is easy to confuse with combustion. Small molecules that are easy to vaporize are called volatile. Nonvolatile substances are composed of large molecules. The latent heat of vaporization for a nonvolatile compound may exceed the energy needed to cause chemical change in the molecules of interest. Heating causes melting followed by **charring**. Combustion involves cleavage of C–C bonds and recombination usually with oxygen (oxidation). Smoke is formed from a combination of charred material (smoke particles) and gaseous compounds (fumes) formed from the thermally decomposing solid.

## References

1. Stevens, M., *Polymer Chemistry. An Introduction*, Oxford University Press, New York, 1999.
2. Rees, D. A. and Smith, J. C., Polysaccharide conformation. VIII. Tests of energy functions by Monte Carlo calculations for monosaccharides, *J. Chem. Soc. Perkin II*, 830–831, 1975.
3. Rees, D. A. and Smith, J. C., Polysaccharide conformation. IX. Monte Carlo calculations of conformational energies for disaccharides and comparison with experiment, *J. Chem. Soc. Perkin II*, 836–841, 1975.
4. McNeil, M., Darviil, A. G., Man, P., Franzen, L.-E., and Albersheim, P., Structural analysis of complex carbohydrates using high-performance liquid chromatography, gas chromatography and mass spectrometry, *Methods Enzymol.* 83, 3–45, 1982.
5. Aspinal, G. O., Chemical characterization and structure determination of polysaccharides, in *The Polysaccharides*, Vol. 1, Aspinal, G. O., Ed., Academic Press, New York, 36, 1982.
6. Launay, B., Doublier, J. L., and Cuvelier, G., Flow properties of aqueous solutions and dispersions of polysaccharides, in *Functional Properties of Food Macromolecules*, Elsevier Applied Science, London, 1–78, 1986.

# 8

# Principles of Food Rheology

## 8.1 INTRODUCTION

### 8.1.1 SIGNIFICANCE AND SCOPE

Rheology is the study of the deformation of materials under stress. Rheological studies provide information concerned with the **elasticity** of solids or the **viscosity** of liquids. Food materials, being neither wholly liquids nor solids, exhibit **viscoelasticity**, i.e., properties intermediate between those for solids and liquids. The rheology of food materials is describable in terms of elastic deformation, viscous flow, and plastic flow.[1] Rheological properties translate into everyday terms such as mouthfeel, smoothness, graininess, thickness, firmness, hardness, brittleness, crispiness, extensibility, and general handling properties.

The **texture** of foodstuffs is related to their perceived consistency: hard versus soft, crispy or not, smooth versus lumpy, flows or globs, stands or flops. Texture is determined by the response of food material to applied forces. According to Matuszek[2] "the energy deposited and the resulting forces ... expose their effects at any level in the hierarchy of food structure, i.e., from the molecular level to the formation of phases, networks, aggregates, cells, and finally the food products themselves." Texture is perceived when food materials are stirred, poured, pumped, stretched, and then, finally, eaten. Rheological characteristics change with variables such as temperature and moisture.

This chapter describes the principle of rheology and some relations between material and rheological properties. Food materials including dilute solutions and so-called soft solids have been subjected to textural and rheological analysis (Tables 8.1 and 8.2).[3]

### 8.1.2 BASIC DEFINITIONS

Rheology is the study of stress and strain. **Stress** ($\sigma$) is the force ($F$) applied per unit area ($A$) of surface. **Strain** ($\gamma$) is the fractional change in length $L$.

$$\sigma = \frac{F}{A}, \tag{8.1}$$

$$\gamma = \frac{\Delta L}{L}. \tag{8.2}$$

Two kinds of external stress may be exerted (Figure 8.1). **Compression** or **tensile stress** is applied at right angles to a surface. The resulting fractional change in length or volume ($dL/L$ or $dV/V$) is termed strain[a]. The second form of stress for materials testing is **shear** or **tangential**

---

[a]In the case of compressional stress, the ratio of strain/stress (($dV/V)/dP$) is called compressibility ($\beta$). We described the relation between compressibility and heat capacity in Chapter 7.

---

**Table 8.1**
**Applications of edible food polymers (hydrocolloids) in foods**

| Application | General actions |
|---|---|
| Viscosity building, gel formation | Stabilization of dispersions, water holding, texture building |
| Encapsulation, film formation | Foam stabilization, crystallization inhibitors, suspension agents |

---

**Table 8.2**
**Examples of products where the rheology is controlled with polymers**

| Food product | Non-food product, or use |
|---|---|
| Bakery products | Adhesives |
| Beverages | Ceramic glazes |
| Chocolate milk | Contact lenses |
| Confectionary products | Cosmetic creams |
| Emulsifiers | Deodorants (carrageen) |
| Encapsulated flavors | Dyes (agar and alginate) |
| Flavor fixatives | Explosives |
| Fried foods | Films |
| Frozen foods | Hair preparations |
| Fruit drinks | Hand lotions |
| Glazes | Inks |
| Gravies | Laxatives |
| Ice cream | Paints |
| Icings | Paper |
| Jams and jellies | Pharmaceuticals |
| Meat products | Printing |
| Milk drinks | Radiography |
| Pie fillings | Textiles |
| Puddings | Tobacco |
| Re-formed fruits | Vaccines |
| Re-formed meat products | |
| Salad dressings | |
| Sauces and soups | |
| Soft drinks | |
| Syrups | |
| Toppings | |
| Yogurt | |

Details adapted from Dea.[3]

---

**stress** applied in a direction parallel to the surface leading to movement of a surface layer in relation to underlying layers. The **strain rate** ($\gamma^*$) is the strain per unit time ($t$), with dimensions of $s^{-1}$:

$$\gamma^* = \frac{\Delta L/\Delta t}{L}. \tag{8.3}$$

The **stiffness modulus** relates stress to strain (deformation) or vice versa. For a solid, the stiffness modulus measures elasticity. When a solid is subjected to compression stress, the size of the resultant strain depends on the constant of proportionality called **bulk modulus** ($K$). Alternatively, shear stress is proportional to the strain rate, with the proportionality constant

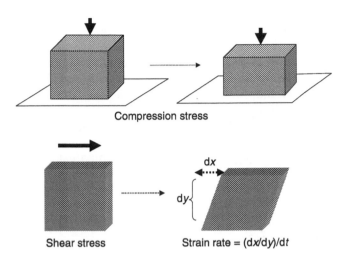

**Figure 8.1** The effect of compression stress (top) or shear stress (bottom) on a block of material.

**Table 8.3**
**Summary of rheological parameters and symbols**

| Parameter (units) | Compression or extension testing | Shear testing |
|---|:---:|:---:|
| Stress (Pa) | $\sigma$ | $\tau$ |
| Strain (dimensionless) | $\varepsilon$ | $\gamma$ |
| Stress/strain (Pa) | $E$ or $K$ | $G$ |
| Strain rate (s$^{-1}$) | $\varepsilon^*$ | $\gamma^*$ |
| Stress/strain rate (Pa s) | $\lambda$ | $\eta$ |

- Solid: Young's modulus ($Y$) = tensile stress ($\sigma$)/strain ($\gamma$)
- Solid: bulk modulus ($K$) = Compression stress ($\sigma$)/strain ($\gamma$)
- Solid: shear modulus ($G$) = Tangential stress ($\tau$)/strain rate ($\gamma^*$)
- Liquid: viscosity ($\eta$) = Tangential stress ($\tau$)/shear rate ($\gamma^*$)

Details adapted from Dickinson.[7]

being the **shear modulus** ($G$). **Young's modulus** ($E$) is the constant relating tensile stress and strain.

For a liquid, the **coefficient of viscosity** ($\eta$) links shear stress and strain rate. These relations are summarized in Table 8.3.[4–6] The dimensions for stress[b] are the same as that for pressure, the pascal (Pa = 1 N/m$^2$) and 1 Pa = 10 dyne/cm$^2$. Viscosity is measured using dimensions of Pa s = 10 poise; 1 mPa s = 1 centipoise (1 cP).

## 8.2 STRESS–STRAIN CURVES

An example of a stress–strain graph for a solid is shown in Figure 8.2. Depending on the technique of study, either stress or strain may be the independent variable (plotted on the x-axis). At small degrees of extension, tensile stress is directly proportional to strain and the system conforms to **Hooke's law**. The material resumes its original dimensions when stress

---

[b]The following SI units will be used: force (N), area = m$^2$, pressure (stress) = force (N)/area (m$^2$) or pascal (Pa). Strain has no units. The strain rate ((dx/dy)/dt) has units of s$^{-1}$.

is removed, i.e., it exhibits **elastic** behavior. With increasing stress or strain, there is deviation from linearity at the limit of elasticity and the material experiences flow. At high values of deformation we reach the limit of plastic flow and the material fractures or breaks. The shaded area of Figure 8.2 shows the energy required for fracture. Other useful rheological indices are listed. Parameters such as the stress or strain at fracture, the yield stress, show some correlations with everyday perceptions of texture.

For **Newtonian** liquids a graph of tangential stress versus shear rate (deformation rate) is a straight line passing through the origin. There are three main types of non-Newtonian behavior leading to deviations from linearity. A **Bingham** fluid behaves (Figure 8.3(a)) like a

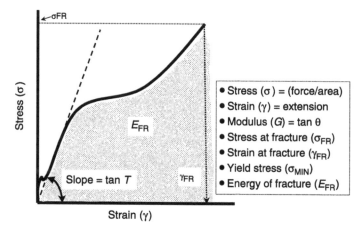

**Figure 8.2**   Stress–strain curve for a rubbery material. Shaded area represents the work needed to cause fracture. The y-axis is sometimes labeled force or load.

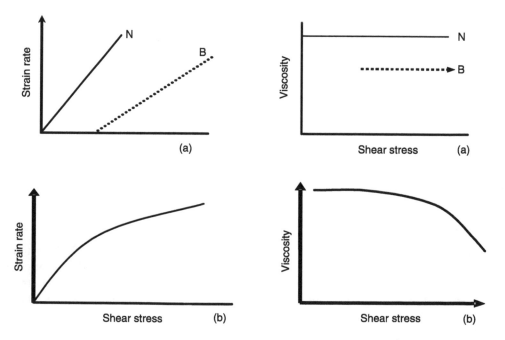

**Figure 8.3**   (a) Comparison of Newtonian (N) and Bingham fluid (B). (b) Pseudoplastic material shear thinning. (c) Dilatant behavior, shear thickening.

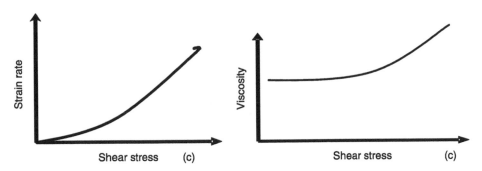

**Figure 8.3** Continued.

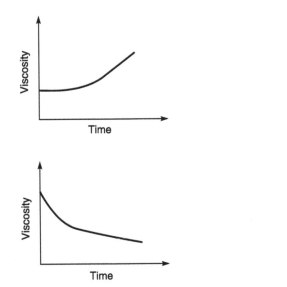

**Figure 8.4** (a) The viscosity of a rheopectic fluid increases with time as it is sheared at a constant rate. (b) Thixotropic behavior: viscosity decreases with time: constant shearing.

solid under static conditions unless the applied force exceeds a certain value called the yield value. Tomato catsup is an example of a Bingham fluid. By comparison, non-Newtonian plastic fluids display either **pseudoplastic** (Figure 8.3(b)) or **dilatant** flow behavior (Figure 8.3(c)). A pseudoplastic fluid is one for which viscosity decreases with increasing shear rate. Many hydrocolloid solutions exhibit this so-called **shear-thinning** behavior. The opposite response is obtained with dilatant fluid which shows **shear thickening**. In this case, viscosity increases with increasing shear rate. A material that shows shear thickening is cornstarch in water. **Rheopectic** and **thixotropic** fluids show time dependent behavior; the apparent viscosity increases and decreases with shearing time (Figure 8.4). A summary of these rheological characteristics is shown in Figure 8.5.

## 8.3   SOLID AND LIQUID FOODS

Dairy products provide examples of foods with liquid (milk), solid (cheese) or intermediate (yogurt) characteristics. We class milk as a liquid because it flows. Cheese is solid-like because it is supports its own weight, also described as free standing. Cheese is also springy to the touch and behaves like an elastic substance. Yogurt flows or globs.

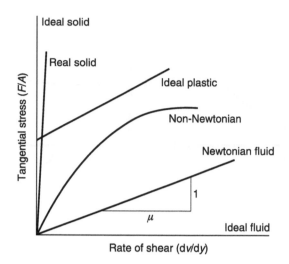

**Figure 8.5** A summary of rheological properties for materials.

Whether a material is classed as a solid or liquid depends on the *observational time frame*. An apparently solid substance will flow under its own weight, given enough time. Lead was once used as roofing on some buildings in Europe. After many centuries, this material flows and distorts. By contrast, most liquids behave as solids over short **observational timescales**. Consider the feel of a pool of water when hit with an open hand, or the sensation of hitting water from a great height.

### 8.3.1 DEBORAH NUMBER

In response to an external stress the molecules in the test material will adjust with a rate of adjustment determined by the relaxation time constant, $\tau''$ (in seconds). For a material subjected to stress for the time $t$, the **Deborah number** (De) is the ratio of the relaxation time to the observation time[c]:

$$De = \frac{\tau''}{t},\tag{8.4}$$

where $De \ll 1.0$ for liquids, $De \gg 1.0$ for solids, and $De \approx 1.0$ for viscoelastic material. By changing the observational time frame, both liquid- and solid-like characteristics may be obtained from the same material. There seems no absolute distinction between solids and liquid materials. Many materials that are of interest to food scientists have $De = 1$ *on the observational timescales accessible to humans.*[7]

## 8.4   MEASURING THE RHEOLOGY OF SOLIDS

### 8.4.1   THE INSTRON TEXTURE ANALYZER

Large-scale deformation measurements, using the Instron analyzer, are simple to take. However, measurements using the classical Instron analyzer are irreproducible and difficult to

---

[c]The relaxation time for a process is simply the inverse of the rate constant ($k$; s$^{-1}$). The relaxation time $\tau$ ($= 1/k$) has dimensions of seconds.

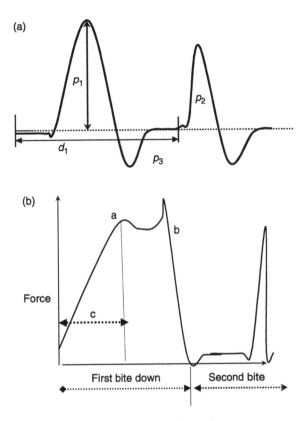

**Figure 8.6** (a) An idealized trace from a General Foods texturometer. $p_1$ = hardness, $p_1/p_2$ = cohesiveness, $p_3$ = adhesiveness. (b) Stress (y-axis) versus strain (x-axis) trace from an Instron texture analyzer; these large deformation tests allow measurement of brittleness = $a$, fracture point = $b$, elasticity = ratio of two peaks, cohesiveness = $c$, and firmness or rigidity = $a/c$.

interpret. Microprocessor controlled texture analyzers are now available with improved reliability and reproducibility. Traces from the common industrial texture analyzers are shown in Figure 8.6. From these graphs it is possible to derive measures for handedness, brittleness, and cohesion. Regardless of the detailed structure and design, rheological instrumentation performs stress–strain measurements. Industrial tests involve large or small deformations that attempt to simulate particular applications. Small-scale deformation measurements using constant stress or oscillating stress are designed not to disrupt the structure under study. The relationship between objective measurements and subjective descriptions of texture is described in Section 8.8.

### 8.4.2 STATIC MEASUREMENT

Two types of static measurements are commonly applied. During a **creep compliance** ($J$) test a fixed magnitude stress is applied and changes in strain (e.g., length, volume) are monitored with time. Alternatively, a **relaxation test** can be applied by monitoring changes in stress with time under a constant applied strain.

### 8.4.3 OSCILLATORY MEASUREMENTS

A small sinusoidal deformation (strain) is applied and the resistive force (stress) generated is monitored. Solids store energy when deformed. Removing the stress results in recoil and recovery of the initial shape due to elasticity. For a perfectly elastic, ideal solid, the maximum stress occurs when the applied deformation is maximum. The strain and stress are said to be in-phase.

For a Newtonian liquid subjected to sinusoidal strain, the stress is maximum when the *instantaneous* strain is zero but the strain rate is maximum. Liquid-like character is measured by the out-of-phase stress. Liquids flow under shear stress to an extent inversely related to their viscosity. Viscoelastic materials show partial solids and liquid-like character. The results of **oscillatory rheology** are the composite modulus ($G^*$). This stiffness index is composed of the **storage modulus** ($G'$), which measures solid-like character and **loss modulus** ($G''$), which shows liquid-like behavior:[8,9]

$$G' = \frac{\text{in-phase stress}}{\text{strain}}, \tag{8.5}$$

$$G'' = \frac{\text{out-of-phase stress}}{\text{strain}}, \tag{8.6}$$

$$G^* = \left(G'^2 + G''^2\right)^{1/2}. \tag{8.7}$$

A **mechanical spectrum** is produced by recording $G'$ and $G''$ as a function of increasing frequency ($\omega$, rad/s) of applied strain. The loss angle ($\delta$) is another sensitive indicator of solid/liquid-like character (Eq. 8.8). The ratio of strain frequency to $G^*$ gives a measure of "dynamic" viscosity, $\eta^*$ (Eq. 8.9).

$$\tan \delta = \frac{G''}{G'}, \tag{8.8}$$

$$\eta^* = G^*/\omega. \tag{8.9}$$

For solid materials we expect $G' > G''$. Both parameters will be independent of strain frequency. For liquids $G'' > G'$ and both parameters increase with oscillation frequency. At high oscillation frequencies, $G''$ and $G'$ cross-over and the system changes from a liquid to solid-like behavior (Table 8.4).

### Table 8.4
### Rheological properties of solids, liquids, and gels

| System | Rheological properties |
|---|---|
| Solid | Stress produces corresponding strain (deformation) |
| | Strain produces stress immediately |
| | Stress and strain are in phase |
| | Removing the stress produces elastic recoil |
| | $G'$ (storage modulus) $> G''$ (loss modulus) |
| | $G'$ and $G''$ are independent of frequency of sinusoidal deformation |
| Liquid | Stress produces flow |
| | Strain lags behind stress |
| | Stress and strain are 90° out of phase |
| | $G'' > G'$ |
| | Modulus increases with frequency |
| Gel | System shows frequency dependence |
| | Liquid-like character at low frequencies |
| | Solid-like character at high frequency |
| | $G'' < G'$ and $G'' > G'$ cross over at high frequency |

---

**Table 8.5**
**Composition of model foods examined by oscillatory rheology**

| Cake batter | % (w/w) | Ice cream mix | % (w/w) |
|---|---|---|---|
| Flour | 27 | Water | 63 |
| Sugar | 27 | Sugar | 12 |
| Whole egg | 35 | Milk solids (non-fat) | 12 |
| Sodium bicarbonate | 1 | Butter | 12 |
| Water | 8 | Emulsifier | 1 |
| Whipping aid | 2 | Air | * |

*Ice cream had 50% gas, i.e., 100% foam overrun after ex-votator.

---

Mechanical spectroscopy was applied to cake batter and ice cream mix prepared according to the formulation shown in Table 8.5.[8] Then 2 ml samples were subjected to oscillatory rheological measurements whilst heating or cooling at a fixed rate. The oscillatory shear frequency was $10 \, s^{-1}$. Parameters $G'$, $G''$ and $\tan \delta$ showed changes at temperatures corresponding to structure formation.

For the cake batter system, $G'$ and $G''$ decreased steadily from 0–80°C owing to the effect of temperature on viscosity but $\tan \delta$ remained constant. Slight changes in $\tan \delta$ occurred at 70°C probably due to starch gelatinization. A large drop in $\tan \delta$ at 80–90°C was ascribed to the denaturation and gelation of egg proteins in the mix. There was a further decrease in the value for $\tan \delta$ at 95–100°C indicative of an increase in rigidity that leveled off at 105–110°C. For ice cream, cooling from −5 to −35°C produced a linear increase in $G'$ from −5 to −20°C and reached max ($10^7$ dyne/cm$^2$) from −25 to −35°C. Interestingly, $G''$ also increased from −5 to −16°C and then decreased. Values for $\tan \delta$ were not reported for ice cream but the data suggests that solidification occurred from −16 to −30°C.[8]

## 8.5 SOLID AND JELLY-LIKE FOODS

### 8.5.1 BULK MODULUS FOR SOLIDS

Solid-like behavior is expected at high Deborah numbers. We can model a solid according to our previous discussion concerning crystalline materials. Each molecule within a crystal lattice vibrates around a fixed point. Adjacent atoms can be compressed only until their electron clouds overlap.[4] The net interaction energy is the result of van der Waals' interactions that reflect the net attraction and repulsion between neutral atoms separated by a distance ($r$). The interaction energy ($E_V$) is expressed by the Leonard–Jones equation:

$$E_V = \frac{B}{r^{12}} - \frac{A}{r^6},$$

(8.10)

where $B$ (=$2.75 \times 10^6$ kcal Å$^{12}$/mole) and $A$ (=1425 kcal Å$^6$/mole) are equation parameters, and $r$ is the effective radius of an atom. Values for $A$ and $B$ are for interactions between two carbon atoms as given by Levitt.[10] Also see Creighton.[11]

From Eq. 8.10 we know that $E_V$ = force ($F$) × distance ($r$), and therefore the force between two atoms is $d(E_V)/dr$. Numerical calculations showing $E_V$, $d(E_V)/dr$ and the stiffness modulus are shown in Figure 8.7. $E_V$ becomes very large for separation distances less than 4 Å ($4 \times 10^{-10}$ m). The interaction force becomes negative (repulsion) and the stiffness modulus increases as adjacent atoms are compressed beyond the optimum separation distance of about 4$r$.

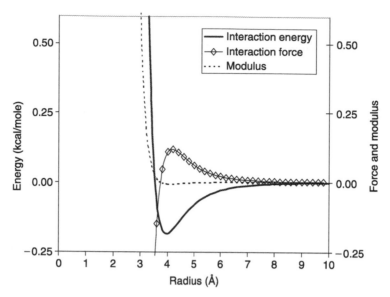

**Figure 8.7**   van der Waals' interaction energy for two carbon atoms with a radius of 1 Å. The interaction force $= \mathrm{d}$ (energy)/d$r$ and the modulus $\propto -1{*}$force/$r^2$. The functions were visualized using Microsoft® Excel.

### 8.5.2   MODULUS OF JELLY-LIKE FOODS

According to the **theory of rubber elasticity**, the stiffness modulus for a jelly-like material is determined by the number density ($n$) of the **virtual springs** or cross-links present. At ambient temperatures thermal kinetic energy ($k_B T$) is stored within the springs:[12]

$$G = n k_B T = \left(\frac{c N_A}{m}\right) k_B T, \tag{8.11}$$

where $G =$ shear modulus, $c =$ the concentration of polymer (g/cm$^3$), $m =$ molecular weight of spring units, $N_A =$ Avogadro's number. The nature of our hypothetical "springs" need not be defined precisely. Eq. 8.11 has been applied to a range of systems including (1) cross-linked elastomers, (2) physical gels, (3) concentrated polymer solutions with transient network interactions, and (4) so-called self-associating polymers.

## 8.6   LIQUID FOODS

### 8.6.1   VISCOSITY

Consider two layers of a liquid each with a surface area, $A$ (cm$^2$) separated by a small distance (d$x$). The topmost layer of liquid is subjected to force, $F$(N) and therefore moves at a velocity $V_1$. The deeper layer of fluid, not subjected directly to the same force, will also move but with a slower velocity ($V_2$). A velocity gradient (d$V$/d$x$) will form (Figure 8.8). The **coefficient of viscosity** ($\eta$) is the proportionality constant connecting applied force (per unit area) and the velocity gradient generated between different layers of liquid (Figure 8.8).

As described before, the strain rate ($\gamma{*}$) has dimensions of per second (s$^{-1}$) whilst tangential stress ($\tau$) is measured as N/m$^2$. This means that the units of viscosity are N.s/m$^2$. According to the CGS system of units, viscosity is measured in units of poise where 1 poise

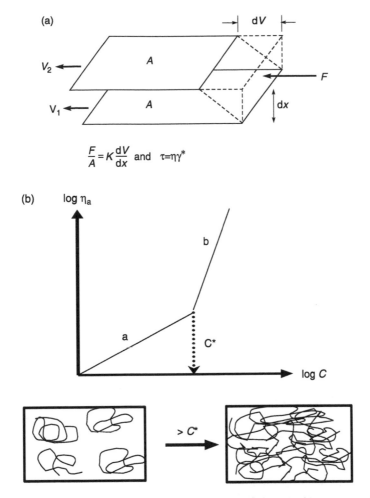

**Figure 8.8** (a) A diagrammatic representation of a liquid undergoing shear stress. (b) The effect of polymer concentration on the viscosity. The two limiting slopes of this graph can be described by a: $\eta_s \approx C^{1.4}$ or b: $\eta_s \approx C^{3.4}$. The transition from a dilute to a semi-dilute solution occurs at the critical polymer concentration ($C^*$).

equals 1 dyne s/cm². A material requiring a shear stress of 1 dyne/cm² to produce a shear rate of s⁻¹ has a viscosity of 1 poise or 100 centipoise.

Dissolved polymers increase solvent viscosity. Placing a molecule between the two layers of fluid in Figure 8.8 reduces the **velocity** gradient (strain rate) generated by a constant applied stress. The newly inserted molecule couples the two layers of liquid together thereby reducing the strain rate ($\gamma^* = dV/dx$). Finally, since $\eta = \tau/\gamma^*$, reducing $\gamma^*$ increases viscosity. This thickening action is achieved with dissolved polymers (Tables 8.1 and 8.2). The consumer expects soups, gravies, pie fillings, and milk drinks to have certain consistencies, pouring characteristics, and mouth feel. Actually, the science behind viscosity generation is fairly simple. *Thickening action is dependent on the size of the dissolved polymer molecule.* As described above, viscosity defines the relationship between strain rate (shear rate) and the perceived stress. The relation between strain rate and viscosity is given in Eq. 8.12:

$$\tau = A + \eta \gamma *^B. \tag{8.12}$$

**Table 8.6**
**Viscosity terms in semi-dilute solution**

| | |
|---|---|
| Relative viscosity | $\eta_{\mathrm{rel}} = \dfrac{\eta_s}{\eta_0} = \dfrac{t_s}{t_0}$ |
| Specific viscosity | $\eta_{\mathrm{sp}} = \eta_{\mathrm{rel}} - 1 = \dfrac{\eta_s - \eta_0}{\eta_0}$ |
| Reduced viscosity | $\eta_{\mathrm{red}} = \dfrac{\eta_{\mathrm{sp}}}{C}$ |
| Intrinsic viscosity | $\eta = \left(\dfrac{\eta_{\mathrm{sp}}}{C}\right)_{C \to 0}$ |

The constants $A$ and $B$ determine the observed behavior (Figure 8.3). Newtonian fluids are those for which $B = 1$ and $A = 0$. For Bingham–Newtonian fluids $B = 1$ and $A \neq 0$. These solutions flow only after a certain "yield stress" is applied. For $B > 1$ there is an apparent thickening with increasing strain rate. Of greater interest are systems with $B < 1$ which exhibit shear thinning (Figures 8.3 to 8.5). The structural basis for shear thinning will be addressed in Section 8.6.6.

Several viscosity terms apply to so-called dilute solutions. **Relative viscosity** is equal to the solution viscosity ($\eta_s$) or flow rate ($t_s$) divided by the solvent viscosity ($\eta_0$) or flow rate ($t_0$), respectively. The **specific viscosity** is the fractional relative viscosity expressed in proportion to the solvent. Adjusting the specific viscosity for concentration gives **reduced viscosity**. To determine **intrinsic viscosity** draw a graph of $\eta_{\mathrm{red}}$ versus concentration and then extrapolate viscosity values to zero concentration. Viscosity terms in semi-dilute solution are given in Table 8.6.

Intrinsic viscosity at infinitely low concentration depends on the properties of individual polymers that are not interacting with neighboring molecules. Intrinsic viscosity depends on the effective volume of a polymer or its radius of gyration (Pg. 120). Another indicator of size is the volume occupied by a given mass—the molecular density. From elementary chemistry we know that density is volume divided by mass ($M$). The effect of polymer dimensions, rigidity, and interaction with the solvent (solvent quality) on intrinsic viscosity is described in the next section[d]. The viscosity of a relatively concentrated (so-called semi-dilute) solution is termed **apparent viscosity** ($\eta_a$).

### 8.6.2 MOLECULAR SIZE AND INTRINSIC VISCOSITY

The relationship between polymer hydrodynamic volume and intrinsic viscosity is described by the **Flory–Fox equation**:[4]

$$\eta = \Phi \frac{R_{\mathrm{ex}}^3}{M}, \tag{8.13}$$

where $\Phi$ is a constant ($3 \times 10^{-24}\,\mathrm{mol}^{-1}$). The expression $R_{\mathrm{ex}}^3 / M$ is volume divided by mass, i.e., density. The effect of polymer molecular weight on the intrinsic viscosity is described

---

[d]The molecular weight is controlled by altering the conditions for polymer synthesis. Alternatively, a large polymer can be degraded into smaller fragments and those fragments separated to give low molecular weight fractions.

**Table 8.7**

**Mark–Houwink parameters for some edible polymers**

| Polymer | Solvent | $v$ | $10^6 K$ | $\eta$ (dl/g) |
|---|---|---|---|---|
| Amylose | Water | 0.68 | 132 | 1.59 |
| Amylose | 0.33 M KCl | 0.50 | 1150 | 1.15 |
| Hydroxyethyl starch | ? | 0.35 | 2910 | 0.37 |
| Alginate | 0.1 M NaCl | 1.0 | 20 | 20.0 |
| Pectin | Water | 0.79 | 11.9 | 11.9 |
| Xanthan | 0.5% NaCl | 0.93 | 24.0 | 24 |

by the Mark–Houwink equation:

$$\eta = KM^v. \tag{8.14}$$

The **Flory–Fox** relation (Eq. 8.13) and **Mark–Houwink** equation (Eq. 8.14) relate polymer size to intrinsic viscosity. The constant $K$ from Eq. 8.14 includes, within it, the size of the monomer subunits or segment size ($L$), their unit mass ($m$), and the characteristic ratio which is a measure of polymer stiffness.

To achieve greater insight about the relations between the Mark–Houwink and Flory–Fox equations and the constants in Table 8.7, we can express the Flory–Fox equation (Eq. 8.13) in terms of the experimentally measurable radius of gyration (Section 7.6.2):

$$\eta = \Phi \frac{6^{3/2} R_g^3}{M}. \tag{8.15a}$$

Next, the characteristic ratio, $C_\infty$ is also written in terms of measurable quantities by recalling, from the random walk model (Pg. 119), that $R = L^2 N$ ($=L^2 M/m$) where $M =$ polymer molecular weight and $m =$ monomer molecular weight. Therefore

$$C_\infty = \frac{R_g^2}{L^2(M/m)}. \tag{8.15b}$$

Finally, Eq. 8.15b is rearranged for $R_g$ followed by substitution into the Flory–Fox equation (Eq. 8.15a) to give

$$\eta = KM^v, \tag{8.15c}$$

$$K = 6\Phi L^3 \left(\frac{C_\infty}{m}\right)^{3/2}. \tag{8.15d}$$

Clearly, the constant $K$ depends on polymer rigidity, unit mass per monomer, and apparent volume ($L^3$).

Values for $K$ and $v$ vary in accordance with polymer–solvent interactions (Table 8.7). For example, $v = 1.0$–1.2 for **free draining coils** where the dissolved polymer shows no interactions with the solvent. By comparison, $v = 0.5$–0.8 for **non-free draining coils** which refers to cases where a strongly hydrated polymer and associated solvent diffuse en masse. Under so-called **theta conditions** a solvent has minimal impact on biopolymer structure. This can be due to

**Table 8.8**
**Values for intrinsic viscosity for some biopolymers**

| Biopolymer | $\eta$ (cm$^3$/g) | Molecular weight (g/mole) |
|---|---|---|
| Polysaccharides | | |
| Amylopectin | 127 | $90 \times 10^6$ |
| Amylose | 81 | $4.9 \times 10^5$ |
| Pectin (LMP) | 185 | $2.5 \times 10^5$ |
| Sodium alginate | 225 and 3100 | $1.2 \times 10^5$ and $1.5 \times 10^6$ |
| Xanthan gum | 5000–8000 | – |
| Globular proteins | | |
| Ribonuclease | 3.3 | $13.7 \times 10^3$ |
| Serum albumin | 3.7 | $66.0 \times 10^3$ |
| Catalase | 3.9 | $250.0 \times 10^3$ |
| Fibrous (rod shaped) proteins | | |
| Collagen | 1150 | $345 \times 10^3$ |
| Myosin | 217 | $493 \times 10^3$ |

Details compiled from Ross-Murphy[1] and Michel et al.[15]

inherent properties of the solvent, due to the addition of salt, or with changing temperature. A further general rule is that $\nu = 0.5$ for random polymers whereas $\nu = 1.8$ for an extended rod-like polymer.[13] Values for intrinsic viscosity for some biopolymers are given in Table 8.8. Further information on this topic can be found in Ross-Murphy[1] and Mitchell.[14]

### 8.6.3 IONIC STRENGTH AND VISCOSITY

The viscosity of a charged biopolymer changes with ionic strength. In a low salt environment, electrostatic repulsion between polymer chain segments leads to an expanded volume, increased rigidity, enhanced solvent binding, higher excluded volume, and increased characteristic ratio and intrinsic viscosity. This so-called **electroviscous** effect depends on (1) the number and location of ionized groups, (2) the average dissociation constant (p$K_a$), and (3) ionic strength. Eq. 8.16 shows the dependence of intrinsic viscosity ($\eta$) on ionic strength ($I$). A graph of $\eta$ versus $1/I^{0.5}$ produces a straight line with gradient $\alpha_1$. The intercept ($\eta_{ref}$) is the apparent viscosity at infinitely high salt concentration:[15]

$$\eta = \eta_{ref} + \frac{\alpha_1}{\sqrt{I}}. \tag{8.16}$$

### 8.6.4 TEMPERATURE AND VISCOSITY

The effect of temperature on viscosity is described by an exponential function (Eq. 8.17) where $\Delta E$ is the activation energy. A graph of $\ln \eta$ versus $1/T$ is linear with a gradient equal to $\Delta E/R$. From Eq. 8.18 we may estimate viscosity at a temperature $T_2$ based on values from another temperature, $T_1$.

$$\eta = A \exp(\Delta E/RT), \tag{8.17}$$

$$\ln \eta_1 - \ln \eta_2 = \frac{\Delta E}{R}\left(\frac{1}{T_1} - \frac{1}{T_2}\right). \tag{8.18}$$

Viscosity increases with *decreasing* temperature. The meaning assigned to $\Delta E$ depends on the model for viscous flow of which there are two. According to the **free-volume model**, flow requires the formation of a free volume larger than the volume of a solute molecule. By comparison, the **activation energy model for viscous flow** states that each solvent molecule is surrounded by a coordination sphere of neighbors much like the ordered arrangement within a crystalline solid. However, unlike a solid the liquid state has each solvent molecule surrounded by an empty coordination site. Flow occurs when solvent molecules jump into the adjacent empty coordination sites. Each jump involves an exchange of surrounding neighbors. In summary, $\Delta E$ for viscous flow can be modeled as the resistance to free volume creation and/or energy needed to displace molecules into empty coordination sites.

### 8.6.5 Effect of Concentration on Zero-Shear Viscosity

Solution viscosity, $\eta_s$, increases with the concentration of dissolved polymer (Eqs 8.19 and 8.20). In general, plotting a graph of log ($\eta_s$) versus log $C$ gives a straight line with a gradient of 1.1 or 3.4 at low and high concentration of polymer, respectively:

$$\eta_s = C^{1.4}, \tag{8.19}$$

$$\eta_s = C^{3.4}. \tag{8.20}$$

The slope change from 1.4 to 3.4 occurs at critical polymer concentration designated $C^*$ (Figure 8.8(b)). At a polymer concentration equal to $C^*$ all the free volume in the solution is taken up by polymer chains.

According to the model proposed by Greassely, entanglement of adjacent polymer chains occurs above $C^*$. A solution for which $C \geq C^*$ is described as **semi-dilute**. The concentration of polymer chains in the bulk solution (grams of polymer dissolved per ml) approaches the density (grams per ml) of individual polymer molecules as one proceeds from a dilute to the semi-dilute regime. Intuitively, we expect that $C^*$ is inversely related to the polymer hydrodynamic volume. The condition for coil entanglement is $C \geq C^*$ where

$$C^* \cong \frac{M}{N_A v^*}. \tag{8.21}$$

The term $v^*$ $(= 4/3\pi R_g^3)$ is the hydrodynamic volume per molecule and $N_A$ is Avogadro's number $(6.023 \times 10^{23}$ molecules per mole). $N_A v^*$ corresponds to the net hydrodynamic volume $(cm^3$ per mole) of polymers. Finally, multiplying $C^*$ $(g/cm^3)$ and $\eta^*(cm^3/g)$ yields a dimensionless number $(S)$ that shows the degree of polymer overlap:

$$S = C^* \eta^*. \tag{8.22}$$

### 8.6.6 Effect of Shear Rate on Viscosity

The viscosity of solutions that exhibit **shear thinning** can be described (Figure 8.9) by the empirical equation

$$\eta_a = \frac{\eta_z}{(1 + 18\gamma^*)^{0.76}} \tag{8.23}$$

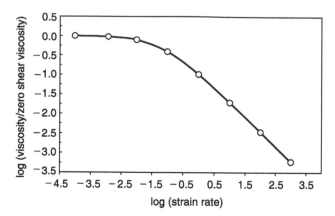

**Figure 8.9**   Shear thinning as described by Eq. 8.23.

where $\eta_z$ is the **zero shear viscosity**. Many polymer solutions exhibit such behavior though the exact position of the viscosity–strain rate curve on the $y$-axis differs in Figure 8.9. The value for $\eta_z$ and the strain rate necessary to reduce $\eta_a$ by 90% (termed $\gamma^*_{0.01}$) are useful indices for the degree of thickening action expected from different polymers. Shear thinning as described by Eq. 8.23 is shown in Figure 8.9.

Shear thinning is associated with semi-dilute and/or concentrated polymer solutions for which there is coil–coil entanglement. The arrangement of polymers in a semi-dilute solution is as shown in the right-hand box in Figure 8.8(b). When such a solution is stirred at a sufficiently fast strain rate ($s^{-1}$) the entangled polymers become disentangled at a rate higher than the rate of re-entanglement. Consequently, there is a decrease in the apparent solution viscosity. For xanthan, an essentially rigid rod-like biopolymer, shear thinning is attributed to the alignment of polymers within the flow stream produced by stirring.

## 8.7   MEASURING VISCOSITY: THE CONE VISCOMETER

The cone plate viscometer is frequently used to measure the viscosity of thick solutions and semi-solid material. Essentially it consists of an inverted cone with an apex lightly resting on a steel plate. The rotating cone makes an angle ($\theta$) with the flattened plate. For a cone rotating at an angular velocity of $\Omega$ (rad/s), the applied stress ($\tau$) and strain rate ($\gamma^*$) are described by

$$\tau = \frac{3M}{2\pi R^3} \quad \text{and} \quad \gamma* = \frac{\Omega}{\theta}, \tag{8.24}$$

where $M$ is the perceived torque (N/m). The dynamic viscosity ($\eta^*$) is simply the ratio of the applied stress to the observed strain rate (or so-called shear rate). In practice, cone viscometer design allows the measurement of torque whilst the cone velocity is varied:

$$\eta^* = \frac{M\kappa}{\Omega}, \tag{8.25}$$

where $\kappa = 3\theta/(2\pi R^3)$ is the instrument constant that depends on the instrument design. Results typical of those obtained using a cone viscometer are shown in Figure 8.4.

---

**Table 8.9**
**Relations between subjective and quantitative rheological indices**

| Subjective measure | Semi-quantitative rheological indices |
|---|---|
| **Solutions** | |
| Thickness | ≈ Viscosity between tongue and palate |
| Smoothness | ≈ 1/(friction force between tongue and palate) |
| Slipperiness | ≈ 1/(frictional + viscous forces between tongue and palate) |
| **Gels** | |
| Chewiness | Firmness × cohesiveness × elasticity |
| Gumminess | Firmness × cohesiveness |
| Stiffness | Strain/stress |
| **Gels and solutions** | |
| Creaminess | $0.1 \times (\text{mushiness})^{0.86} \times (\text{cohesiveness})^{0.99}$ |
| Creaminess | $0.69(\text{cohesiveness}) - 0.85$ |
| Hardness | $3.16 + 0.25(\text{crispiness}) + 0.76(\text{toughness})$ |
| Gumminess | $1.71 + 0.60(\text{cohesiveness}) + 0.20(\text{chewiness})$ |
| Viscosity | $28.56 + 0.48(\text{adhesiveness})$ |

---

## 8.8 SENSORY EVALUATION OF TEXTURE

The relations between objective rheological measurements and **subjective sensory evaluations** are of some interest. Subjective descriptors for liquids include terms like viscous, slimy, smooth, slippery, and/or sticky. Solid foods are described as firm, grainy, brittle, chewy, smooth, adhesive, cohesive or fracturable. The issue is to match such terms with instrumental measurements, taking account of the wide range of stress–strain responses possible. A further complication arises from the possibility that sensory perceptions may be specific to each subject and/or modified by culture and learned behavior.[16] Some suggested relations between subjective and quantitative rheological indices are summarized in Table 8.9.

Polymer solutions that exhibit Newtonian behavior are perceived as slimy.[17–19] Recall that Eq. 8.13 describes stress vs strain rate relations for liquids and that the power term, $B$ takes on values of $<1$ (shear thinning), 1 (Newtonian response) or $>1$ (shear thickening). Pseudoplastic behavior (shear thinning) was scored as non-slimy by panelists. Perceived sliminess was linearly related to $B$:

$$\text{sliminess} = \frac{(B - \text{constant})}{m}, \tag{8.26}$$

where $m$ is a proportionality value, and $B$ is related to shear stress–strain rate behavior of liquids. Subjective score for **creaminess** was optimum for materials where $B = 0.5$. Materials for which $A \neq 0$ (cf. Eq. 8.13) have significant yield stress and a tendency to stick to the walls of containers when poured. For a thorough review of this subject the reader is referred to Sherman.[16]

## References

1. Ross-Murphy, S. B., Rheological methods, in *Physical Techniques for the Study of Food Biopolymers*, Blackie Academic & Professional, New York, 343–392, 1994.
2. Matuszek, T., Rheological properties of food systems, in *Chemical and Functional Properties of Food Components* (2nd Edition), Sikorski, Z. E., Ed., CRC Press, Washington, D.C., 179–204, 2002.

3. Dea, I. C. M., Conformational origins of polysaccharide solution and gel properties, in *Industrial Gums: Polysaccharides and Their Derivatives*, Whistler, R. L. and BeMiller, J. N., Eds, IRL Press, Oxford, 21–52, 1993.

4. Goodwin, J. W. and Hughes, R. W., *Rheology for Chemists: An Introduction*, Royal Society of Chemistry, Cambridge, UK, 2000.

5. Anandha Rao, M., *Rheology of Fluid and Semi-solid Foods. Principles and Applications*, Aspen Publishers, Gaithersburg, MD, 1999.

6. Steffe, J. F., *Rheological Methods in Food Process Engineering* (2nd edition), Freeman Press, East Lansing, MI, 1996.

7. Dickinson, E., *An Introduction to Food Colloids*, Oxford University Press, 51–78, 1992.

8. Dea, I. C. M., Richardson, R. K., and Ross-Murphy, S. B., Characterization of rheological changes during the processing of food materials, I, *Gums & Stabilizers* 2, 357–374, 1984.

9. Oakenfull, D., Pearce, J., and Burley, R. W., Protein gelation, in *Food Proteins and Their Applications*, Damodaran, S. and Paraf, A., Eds, Marcel Dekker, New York, 111–167, 1997.

10. Levitt, M., Energy refinement of hen egg-white lyzozyme, *J. Mol. Biol.* 82, 393–420.

11. Creighton, T. E., *Proteins: Structure and Molecular Properties*, Freeman, W. H., Ed., New York, 137–139, 1984.

12. Flory, P. J., *Principles of Polymer Chemistry*, Cornell University Press, Ithaca, 1953.

13. Launay, B., Doublier, J. L., and Cuvelier, G., Flow properties of aqueous solutions and dispersions of polysaccharides, in *Functional Properties of Food Macromolecules*, Mitchell, J. R. and Ledward, D. A., Eds, Elsevier Applied Science, London, 1–78, 1986.

14. Mitchell, J. R., Rheology of polysaccharide solutions and gels, in *Polysaccharides in Foods*, Blanshard J. M. V. and Mitchell, J. R., Eds, Butterworths, London, 1979.

15. Michel, F., Doublier, J. L., and Thaibault, J. F., Investigations on high-methoxyl pectins by potentiometry and viscometry, *Prog. Food Nutri. Biosci.* 6, 367–372, 1982.

16. Sherman, P., Hydrocolloid solutions and gels. Sensory evaluation of some textural characteristics and their dependence on rheological properties, *Gums & Stabilizers* 6, 269–284, 1982.

17. Szczesniak, A. S. and Skinner, E. Z., Meaning of texture words to the consumer, *J. Texture Stud.*. 4, 378–384, 1973.

18. Szczesniak, A. S., Rheological basis for selecting hydrocolloids for specific applications, *Gums Stabilizers* 3, 311–323, 1986.

19. Szczesniak, A. S., Textural perceptions and food quality, *J. Food Qual.* 14, 75–85, 1991.

# 9

# Chemistry of Nonenzymic Oxidations

## 9.1 INTRODUCTION

Oxidation and browning are the two chemical routes for food deterioration. **Nonenzymic oxidation** (NEO) reactions are discussed in this chapter. High temperature processes such as pasteurization and UHT treatment increase the rate of NEO. Refrigeration or freezing does not hinder oxidation because the solubility of oxygen in aqueous solution increases at low temperatures. Understanding the chemistry of food deterioration is important if we are to develop effective interventions for such processes. The characteristics of food oxidation (Section 9.1) and redox potentials (Section 9.2) are presented at the start of this chapter. **Reactive oxygen species**, the immediate promoters of oxidation, are introduced in Section 9.3. Nonenzymic **catalysts** for oxidation are described in Section 9.4 followed by oxidation inhibitors, **antioxidants**, in Section 9.5. The final section on NEO (Section 9.6) covers applied aspects of the topic related to some real food systems.

### 9.1.1 OXIDATION

Oxidation is defined as the addition of oxygen or removal of hydrogen from a molecule. A great many oxidation reactions do not involve oxygen directly. Nevertheless, it will be useful to start with the focus on oxygen, this being the most well-known food oxidant. The atmospheric concentration of oxygen is 20% co-existing with 75% nitrogen. A liter of water dissolves about 8 mg of oxygen equivalent to 0.25 mM. Therefore, most fresh foods (80–90% water) have no more than 8 ppm[a] oxygen. The solubility of oxygen is higher in oils as compared with aqueous media.

For non-packaged foods, oxygen exposure is related to the **surface area** available for gaseous diffusion. Smaller food pieces expose larger surface area for oxygen exchange. The rate of oxygen **removal** by **enzymatic** and nonenzymic processes is another consideration as is the presence of **antioxidants** some of which combine directly with oxygen. For packaged foods, oxygen exposure is also controlled by **headspace volume**, or total volume of **void spaces** within the package, the composition of gases within the headspace, and the **permeability** of the packaging to oxygen.

### 9.1.2 SIGNIFICANCE IN FOODS

Oxidation affects most food components, including the **microconstituents**—colors, flavor compounds, vitamins, and minerals. The **macroconstituents**, carbohydrate, lipids, and proteins are also susceptible to oxidation. Most **quality attributes** are affected by oxidation including

---

[a] 1 ppm (part per million) of X = 1 mg (of X) dispersed in 1 kg of food. This classical measure of concentration remains popular and is analogous to 1 ppt (one part thousand) or the percent (one part per hundred).

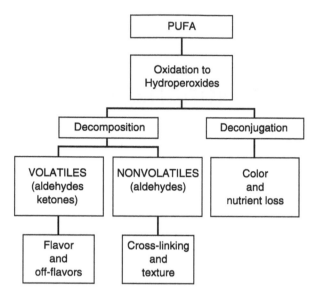

**Figure 9.1**  The effect of oxidation reactions in foods.

sensory properties (flavor, i.e., taste, odor or smell, texture, visual appeal), nutritional properties, shelf-life, and safety. Recent advances in nutritional chemistry show that oxidation is a possible contributor for aging and **degenerative disease**. Oxidation processes, and mechanisms described for simple foods, also impact on diet and health.

**Bleaching** of pigments, destruction of **vitamins**, formation of **off-flavors**, and cross-linking are some of the well-known results of food oxidation. Oxidation destroys conjugated double bonds leading to the loss of color or vitamin potency. Reactions between poly-unsaturated fatty acids and oxygen form **hydroperoxides** (ROOH). These can degrade to lower molecular weight products (ketones, aldehydes, alcohols) that impart distinctive off-flavors.[1] Oxidation has a huge impact on lipids, i.e., *those materials from biological sources with significant solubility in organic solvents (Chapter 5). Lipids include some colors, flavors, and fat soluble vitamins.* The effect of oxidation reactions in foods is summarised in Figure 9.1.

The red **color of fresh meat** is due to oxygenated myoglobin. Oxymyoglobin in which heme iron is bound to oxygen is bright red. Reduced myoglobin is dull-brick color. Some **aldehydic products of oxidation** can encourage cross-linking reactions between proteins and lipids with possible consequences for food texture.

## 9.2  CHARACTERISTICS OF FOOD OXIDATION

The study of oxidation requires that we measure trace amounts of products at concentrations of ppm. Moreover, foods contain a cornucopia of reducing and oxidizing agents. Results obtained with idealized solutions may not transfer readily to complex foods. The *modus operandi* of food oxidation is still not fully understood. However, most oxidation processes involve 1-electron transfer reactions (Table 9.1)[b].

---

[b] Single electron transfer reactions form a distinct branch of chemistry. These reactions are distinct from the breaking and forming of covalent bonds or ionic bonds, which involve the sharing and/or transfer of pairs of electrons. Heterolytic cleavage of covalent bonds to form radicals is not considered oxidation.

**Table 9.1**

**Characteristics of oxidation reactions in foods**

- Frequently involve free radicals with short lifetimes in water and relatively longer lifespan in nonpolar (oily) or gaseous media
- Involve 1-electron transfer, i.e., classical reduction–oxidation (redox) reactions in aqueous solution
- Catalyzed by enzymatic or nonenzymatic agents, including free metal ions, heme, heme proteins, metallo proteins, and flavoproteins
- Relatively nonspecific: highly reactive free radicals attack most molecules
- Shows a lag phase followed by a rapid onset; the time between apparent onset and total product failure is short
- Inhibited by free radical scavengers, chelators or redox compounds
- Requires oxygen or an activated oxygen species
- Hydrogen peroxide transforms most reducing agents into pro-oxidants
- Initiated by light, high temperatures or irradiation

Oxidation is exemplified by dehydrogenation of RH to form the oxidized species $R_{OX}$. Another form of oxidation is the removal of an electron from RH by a **free radical** such as the hydroxyl radical ($OH^\bullet$).

$$RH \to R(ox) + 1e + H^+ \tag{9.1}$$

$$RH + OH^\bullet \to R^\bullet + H\text{–}O\text{–}H \tag{9.2}$$

In Eqs. 9.1 and 9.2 the oxidizing agent acts as an electron acceptor. The reducing agent is an electron donor. Oxidation may also occur via the addition of oxygen to RH to form a hydroperoxide, ROOH.

The kinetics of free radical reactions was described previously (Chapter 7) in terms of the rate of **initiation** and **termination** (Eqs. 9.3a, 9.3b and 9.3c). Oxidation is predominantly a 1-electron process.

$$\text{Initiation: } In \xrightarrow{k_i} In^* \tag{9.3a}$$

$$\text{Initiation: } In^* + M \xrightarrow{k_i} In + M^* \tag{9.3b}$$

$$\text{Termination: } M^*_{(i+i)} + M^* \to M_{(i+2)} \tag{9.3c}$$

$$R_p = k_p \phi [M][In]^{0.5} \tag{9.4}$$

The rate of free radical *oxidation* reactions (Eq. 9.4) depends on (1) the concentration of catalyst or *initiator* ([*In*]), and (2) the concentration of the reacting *species* ([*M*]). This explains why polyunsaturated fatty acids oxidize more readily than their monounsaturated counterparts. **Initiation** and **termination** provide much of the rationale for studying food oxidation. In this chapter we focus on the mechanisms of initiation and the role of oxidation catalyst (Sections 9.3 and 9.4). Later there is emphasis on the termination process and the role of antioxidants (Section 9.5).

## 9.2.1 REDOX POTENTIALS

Compounds can be arranged in the order of their tendency to receive or accept electrons. The redox scale shows a list of **half-cell reactions**, so-called because two half-cell reactions form electrochemical cells (batteries). By convention, a half-cell (Eq. 9.5) is always written as a reduction process. Thus, $X^{N+}$ combines with one electron to form $X^{(N-1)}$. The associated hydrogen ion is not shown. In a reduction process, the oxidation number for compound X decreases from $N^+$ to $(N-1)^+$.

$$X^{N+} + e \rightarrow X^{(N-1)} \tag{9.5}$$

A much-discussed half-cell is formed by inserting a copper wire into a molar solution of copper sulfate (Figure 9.2). As the copper wire corrodes, $Cu^{2+}$ ions dissolve leaving a build-up of negative charge (electrons) on the wire surface. The half-cell potential ($E$) is measured by hooking the half-cell to a reference half-cell (not shown) via a high resistance millivoltmeter to avoid current flow. A common reference half-cell used for redox potential measurements is the **standard hydrogen electrode** (SHE), which has $E_{SHE}$ value set arbitrarily to 0 V. Table 9.2 shows the redox potentials for some foods. To measure $E^{o\prime}$ a clean platinum rod is inserted into the food and this is connected to a reference half-cell. The potential generated at the platinum surface is due to a reversible redox reaction between adsorbed and dissolved species. Values for some redox potentials are listed in Table 9.2.

Table 9.3 shows the redox potentials for some pure compounds. Half-cells listed in the top half are for oxidizing agents as indicated by the positive $E^\circ$ values recorded when these cells

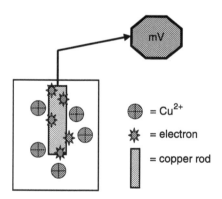

**Figure 9.2**  A schematic diagram of a half-cell for copper sulfate. The millivolt meter (mV) reads the redox potential for the $Cu^{2+} + 2e \rightarrow Cu(solid)$ when the entire half-cell is connected to a reference half-cell (hydrogen electrode).

**Table 9.2**

**Redox potential values for some foods**

| Food | $E^\circ(V)$ | pH |
|---|---|---|
| Pears | 0.439 | 4.2 |
| Grapes | 0.409 | 3.9 |
| Lemons | 0.382 | 2.0 |
| Meat (minced, fresh) | 0.225 | 5.9 |
| Myoglobin | 0.50 | 6.8 |
| Potato tubers | −0.150 | 6.0 |
| Meat (raw) post-rigor | −0.200 | 5.7 |
| Wheat (whole grain) | −0.340 | 6.0 |

**Table 9.3**
**Half-cell reactions and redox potential values for some anions**

| Half-cell reaction | $E^{\circ}$(Volt) | $E'$(pH 7) |
|---|---|---|
| 1. $F + H^+ + e \rightarrow HF$ | 3.06 | 2.647 |
| 2. $OH^{\bullet} + e + H^+ \rightarrow OH^-$ | **2.81** | **2.395** |
| 3. $S_2O_8^{2-} + 2e \rightarrow 2SO_4^{2-}$ | 2.00 | 1.586 |
| 4. $H_2O_2 + 2H + 2e \rightarrow 2H_2O$ | 1.776 | 1.362 |
| 5. $BrO_3^- + 6H^+ + 6e \rightarrow Br^- + 2H_2O$ | 1.52 | 1.106 |
| 6. $Mn^{3+} + e \rightarrow Mn^{2+}$ | 1.51 | 1.096 |
| 7. $ClO_3^- + 6H^+ + 6e \rightarrow \frac{1}{2}Cl(g) + 3H_2O$ | 1.47 | 1.056 |
| 8. $Cl_2(g) + 2e \rightarrow 2Cl^-$ | 1.359 | 0.945 |
| 9. $O_2 + 4H^+ + 4e \rightarrow 2H_2O$ | 1.229 | 0.815 |
| 10. $^{\bullet}O_2H + e + H^+ \rightarrow H_2O_2$ | **1.200** | **0.786** |
| 11. $IO^{3-} + 6H^+ + 6e \rightarrow \frac{1}{2}I(g) + 3H_2O$ | 1.178 | 0.764 |
| 12. $Br(aq) + 2e \rightarrow 2Br^-$ | 1.087 | 0.673 |
| 13. $NO_3^- + 3H^+ + 2e \rightarrow HNO_2(g)$ | 1.00 | 0.421 |
| 14. $Fe(CN)_6^{3+} + e \rightarrow Fe(CN)_6^{2+}$ | 0.774 | 0.36 |
| 15. $Fe^{3+} + e \rightarrow Fe^{2+}$ | 0.765 | 0.351 |
| 16. $H_2O_2 + e \rightarrow OH^{\bullet} + OH^-$ | **0.760** | **0.346** |
| 17. quinone $+ 2e + 2H^+ \rightarrow$ quinol | 0.699 | 0.285 |
| 18. $Cyt\text{-}c.Fe_{(ox)}^{3+} + e \rightarrow Cyt\text{-}c.Fe_{(red)}^{2+}$ | 0.664 | 0.250 |
| 19. $DHA + 2e + 2H^+ \rightarrow$ ascorbic acid | 0.474 | 0.060 |
| 20. $Cu^{2+} + 2e \rightarrow Cu^{\circ}$ | 0.337 | −0.077 |
| 21. riboflavin $+ 2H^+ + 2e \rightarrow$ dihydroriboflavin | 0.206 | −0.208 |
| 22. glutathione$_{(ox)} + 2e + 2H^+ \rightarrow$ glutathione$_{(red)}$ | 0.174 | −0.240 |
| 23. $Cu^{2+} + e \rightarrow Cu^+$ | 0.153 | −0.261 |
| 24. $NAD^{2-} + 2e + 2H^+ \rightarrow NADH$ | 0.094 | −0.327 |
| 25. $2H^+ + 2e \rightarrow H_2$ | 0.0 | −0.414 |
| 26. $O_2 + e + H^+ \rightarrow {}^{\bullet}O_2H$ | **−0.05** | **−0.464** |
| 27. $Sn^{2+} + 2e \rightarrow Sn$ | −0.136 | −0.550 |
| 28. $Ni^{2+} + 2e \rightarrow Ni$ | −0.250 | −0.666 |
| 29. $V^{3+} + e \rightarrow V^{2+}$ | −0.275 | −0.689 |
| 30. $Ti^{3+} + e \rightarrow Ti^{2+}$ | −0.37 | −0.784 |

are connected to a SHE. By comparison, a negative $E^{\circ}$ value implies a strongly reducing agent. Many anions are used as processing chemicals: fluoride, hydrogen peroxide, bromate are useful oxidizing and bleaching agents. The metal cations are frequently found in foods as minerals and are also formed by contact surfaces (cans, work surfaces, pots, and pans).

### 9.2.2  THERMODYNAMICS OF REDUCTION–OXIDATION REACTIONS

The net redox potential change for a spontaneous reaction is positive.[3] The maximum work done when 1mole of electrochemical reaction occurs is expressed by Eqs. 9.6 and 9.7:

$$\Delta G^{\circ} = -nFE^{\circ}, \tag{9.6}$$

$$\Delta G^{\circ} = -2.3RT \log K_{eq} \tag{9.7}$$

where $n$ is the number of electrons transferred per atom reacted, $E^{\circ}$ is the electrode potential, F is Faraday's constant (96,487 J/mol/V or coulomb/mol) and $K_{eq}$ (=[product]/[reactants]) is the equilibrium constant for the redox reaction. For a spontaneous reaction, $\Delta G^\circ$ is negative whilst $E^\circ$ is positive. Recall that such thermodynamic considerations, though useful, will not provide information about rates of oxidation.

### 9.2.3  EFFECT OF CONCENTRATION AND PH

Standard $E^\circ$ values cited above assume that all reactants are present at a concentration of 1 M. At other concentrations the redox potential ($E$) is as expressed by the Nernst equation (Eq. 9.8):

$$E = E^\circ + \frac{0.059}{nF}\log\frac{[1]}{[\text{concentration}]}.$$
(9.8)

For those half reactions involving the hydrogen ion ($H^+$) at concentrations other than 1 M (cf. pH $= 0$), $E$ changes with pH according to Eq. 9.9. For biological systems the redox potential at pH 7 ($E'$) is of interest.

$$E' = E^\circ + \frac{0.059}{nF}\log\frac{[1]}{[\text{concentration}]} - 0.059\,\text{pH}.$$
(9.9)

In summary, a range of factors affect the redox potential including (1) the type of redox compounds (i.e., half-reactions present), (2) the ratio of all oxidants and reductions, (3) the pH of the reaction medium, (4) the availability of oxygen and hydrogen peroxide which are ubiquitous oxidants, and (5) presence of microbial activity. A brief comparison of data in Tables 9.2 and 9.3 suggests that whole wheat has a reducing environment. It might be theorized that this is due to the high concentrations of **glutathione** within wheat grains. The subtleties concerning $E^\circ$ values are worth bearing in mind as we consider the redox properties of various food components later.

## 9.3  REACTIVE OXYGEN SPECIES

Atmospheric oxygen is relatively unreactive. Food compounds react more readily with oxygen after this undergoes activation. In this section we describe the most important activated oxygen species and their involvement in food oxidation. Recall from cell biology that glucose is metabolized in the presence of oxygen to carbon dioxide and water. The respiration process is accompanied by the transfer of four electrons to oxygen, forming water. Figure 9.3 shows the oxygen intermediates formed during the successive reduction of oxygen by a series of 1-electron transfer processes.

Two activated oxygen species are directly involved in food oxidation: (1) the **superoxide radical**, also called the peroxyl radical, and (2) the **hydroxyl radical** ($E^\circ$ values show that the

| Oxygen | Superoxide radical ($^\circ O_2$) + 1 ep | Peroxide + 1 e | Hydroxyl radical ($^\circ OH$) + 1 e | Water |
|---|---|---|---|---|
| $E^\circ = -0.05$ V | $E^\circ = 1.2$ V | $E^\circ = 0.76$ V | $E^\circ = 2.81$ V | |

**Figure 9.3**  Stages in the 4-electron reduction of molecular oxygen to form water.

hydroxyl radical is the most oxidizing). From Table 9.3, we should expect that only elements like Sn, Ti, Ni, and V could transfer electrons directly to molecular oxygen thereby forming a superoxide ion. Moreover, this process would be unlikely without the aid of a catalyst (see below). Now consider the position of hydrogen peroxide in Table 9.3. Clearly, there are many more compounds below hydrogen peroxide which can transform this to the hydroxyl radical.

### 9.3.1 SINGLET AND TRIPLET OXYGEN

Atmospheric oxygen is unreactive because it is in the **ground state**. To understand what this means, we must see how electrons are arranged within molecular orbitals for the $O_2$ molecule (see Chapter 1). There are 16 electrons for the $O_2$ molecule arranged as shown in Figure 9.4.

In a given molecule each electron has a positive spin ($S = +1/2$) or negative spin ($S = -1/2$) spin. The total spin is expressed by the formula $2S + 1$. In the case where there are parallel electrons, the resulting total spin is $2(0.5 + 0.5) + 1 = 3$ (triplet state). If the two electrons have opposite spins the total spin is expressed by $2(-0.5 + 0.5) + 1 = 1$ (singlet state). Notice the presence of two unpaired electrons in the $\pi*2p$ (antibonding) orbital of atmospheric oxygen (Figure 9.4). The low reactivity of triplet oxygen can be explained using the **principle of spin conservation**. According to this idea chemical reactions occur between molecules with the same *net spin*. Singlet molecules react with singlet molecules. Look at the alternative arrangements of $\pi*2p$ electrons for oxygen shown in Table 9.4. The probability that an incoming electron will have an *opposite* spin (and avoids repulsion by an electron with the same spin) increases with singlet oxygen.

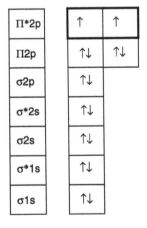

**Figure 9.4**   The arrangement of electrons in molecular orbitals for $O_2$. Alternative electron arrangements in the $\pi*2p$ orbitals of oxygen and related species are shown in Table 9.4.

**Table 9.4**
**Arrangements of electrons in molecular oxygen 2p antibonding orbitals**

| State | $\pi$ orbital (a) | $\pi$ orbital (b) | Energy difference (kJ/mol) |
|---|---|---|---|
| Singlet ($^1\Sigma$) | $\downarrow$ | $\uparrow$ | 155 |
| Singlet ($^1\Delta$) | $\downarrow\uparrow$ | | 92 |
| Triplet ($^3O_2$) | $\uparrow$ | $\uparrow$ | 0 |

The triplet form of $O_2$ is relatively stable because parallel unpaired electrons reside in separate orbitals with low mutual repulsion or attraction (Table 9.4). Compared to atmospheric oxygen, singlet oxygen is more energetic by $92 - 155\,kJ/mol$. Transformation of triplet oxygen to the singlet species increases its reactivity.

### 9.3.2 SUPEROXIDE RADICAL

The **superoxide radical** ($^\bullet O_2$) is another reactive oxy-radical formed as a byproduct of electron transport processes taking place during aerobic respiration. Electrons harvested from the **Krebs cycle** are transferred to the electron transport pathway located in the inner membrane of mitochondria. The succession of 1-electron reactions begins with NADH (the reduced form of nicotinamide adenine dinucleotide) and ends with the transformation of oxygen to water. During this process superoxide radicals may form. Away from energy metabolism, the superoxide radical also forms when oxygen reacts with transition metals. Elements such as $Fe^{2+}$ and $Cu^+$, which are able to sustain multiple oxidation states ($Fe^{3+}$, $Fe^{4+}$, $Cu^{2+}$ etc), will transfer electrons to oxygen. Closely related to superoxide radical, the **peroxyl** radical (OOR) forms when a radical hydrocarbon reacts with oxygen ($^\bullet R + O_2 \rightarrow {}^\bullet OOR$). One route for superoxide control is via the enzyme superoxide dismutase (SOD). This lowers the level of harmful superoxide species by catalyzing a simultaneous reduction and oxidation (dismutation) reaction forming hydrogen peroxide and oxygen (Eq. 9.3). From Figure 9.3, it can be seen that hydrogen peroxide is potentially less damaging than the superoxide radical.

$$^\bullet O_2 + {}^\bullet O_2 + 2H^+ \rightarrow H_2O_2 + O_2 \tag{9.10}$$

SOD works in conjunction with **catalase** and **glutathione reductase**, which further catalyze the breakdown of hydroperoxides (Chapter 12).

### 9.3.3 PEROXIDES AND HYDROPEROXIDES

At least four sources of $H_2O_2$ and hydroperoxides[c] are found within living cells:

(1) *Oxidases.* These enzymes produce hydrogen peroxide as a co-product. Glucose oxidase catalyses the reaction glucose $+ O_2 \rightarrow$ gluconic acid $+ H_2O_2$. Examples of other oxidases include amino acid oxidase, from bacteria, and xanthine oxidase in meat. A significant number of oxidases contain FAD (flavin adenine dinucleotide) as a coenzyme.

(2) *Riboflavin*, the functional component of FAD, is a yellow vitamin found in eggs, wheat, and milk, as well as in meat. Riboflavin is a strong reducing agent capable of reducing oxygen: oxygen $\rightarrow$ superoxide radical $\rightarrow$ hydrogen peroxide (The action of riboflavin as sensitizer is described in Section 9.3.5)

(3) Leakage of a *reactive oxygen species* (including hydrogen peroxide) may occur during the 4-electron reduction of oxygen to water along the electron transport pathway in mitochondria.

(4) *Hydrogen peroxide* is a product of the reaction catalyzed by SOD (Eq. 9.10), which protects cells from oxidative damage. Enzymatic oxidation of polyunsaturated fatty acid (PUFA) leads to hydroperoxides (ROOH). We should not forget that these possess all the chemical attributes of hydrogen peroxide (see below).

---

[c]Hydrogen peroxide is a member of the general family of hydroperoxides R–O–O–H where R is a substituent group such as hydrogen, alkyl or fatty acid.

Large quantities of hydrogen peroxide are potentially available from photosynthesis. The first step in the fixation of carbon dioxide is the combination of $CO_2$ with a 5-carbon sugar; the reaction is catalyzed by a low efficiency enzyme, RUBISCO (ribulose-1,5-bisphosphate carboxylase), which is found in all green leaves and thought to be the most abundant protein in the biosphere. A sizeable amount (25%) of carbon fixation by RUBISCO leads to a waste product called glycolic acid (hydroxyacetic acid). This is metabolized to glycoxylic acid by an oxidase (glycolic acid oxidase) inside cell organelles called **peroxisomes**. This pathway and the associated consumption of oxygen form the process of **photorespiration**. The large quantity of hydrogen peroxide is formed by glycolic acid oxidase and is then broken down by the enzyme catalase, which is found in high concentrations within peroxisomes[g]. Hydrogen peroxide ($E° = 0.76$) is considered a modest oxidizing agent. However, this is readily converted to a hydroxyl radical by the action of prooxidants (see below).

### 9.3.4  HYDROXYL RADICALS

A hydroxyl radical (OH$^\bullet$) forms when peroxide is reduced in a 1-electron reaction (Eq. 9.10). If the reducing agent is a metal ion the process is called the **Fenton reaction**. The hydroxyl radical ($E° = 2.81$ V; Table 9.3) is a strongly oxidizing species.

$$M^{(n+)} + H_2O_2 \rightarrow OH^\bullet + OH^-(\text{alkali}) + M^{(n+1)} \qquad (9.11)$$

The pro-oxidant species $M^{n+}$ involved in the Fenton reaction is generally a transition metal, e.g., $Fe^{2+}$, $Cu^+$, $Zn^{2+}$ or $Mn^{2+}$ (cf. Chapter 5, Section 5.2 on the use of the Fenton reagent as a polymerization catalyst). Hydroxyl radicals also form via the **Harber–Weiss** *reaction* (Eq. 9.12). This reaction is favored thermodynamically since the underlying two half-reactions (Eqs. 9.12a and 9.12b) have a net redox potential change of 0.76 V.

$$H_2O_2 + {}^\bullet O_2 \rightarrow OH^\bullet + OH^-(\text{alkali}) + O_2 \qquad (9.12)$$
$$H_2O_2 + e \rightarrow OH^\bullet + OH^- (E = 0.71 \text{ V}) \qquad (9.12a)$$
$${}^\bullet O_2 \rightarrow O_2 + e \ (E = 0.05 \text{ V}) \qquad (9.12b)$$

### 9.3.5  SINGLET OXYGEN FORMATION BY PHOTOSENSITIZATION

Reactive oxygen species are also produced by the action of light. The process called photosensitization requires a light-sensitive pigment such as riboflavin. Illumination excites the sensitizer from a low energy triplet state into the singlet state. Electromagnetic energy is then transferred to oxygen, which is converted from the triplet to a singlet state (Table 9.4). No electron transfer occurs from the sensitizer (Sen) to oxygen.

$${}^3\text{Sen} \xrightarrow{h\nu} {}^1\text{Sen}^* + {}^3O_2 \rightarrow {}^3\text{Sen} + {}^1O_2. \qquad (9.13)$$

A closely allied process of activation occurs when oxygen is heated. The formation of singlet oxygen requires about 92–155 kJ/mole energy whereas the energy cost is 4.82 kJ/mole when oxygen is activated by a 1-electron transfer reaction leading to the superoxide radical.

In summary, reactive oxy-radicals are formed via various biochemical and technological processes: (1) as byproducts of respiration, (2) via the Fenton reaction, (3) via the Harber–Weiss reaction, (4) by thermal activation, and (5) via photosensitization. Notice that the discussions in Section 9.3 are essentially thermodynamic in nature. They consider what reactions *could* occur based on values of $E°$ (Eqs. 9.6 and 9.7). There is no reference to the rate of reaction. In Section 9.4, we consider some variables affecting the rate of food oxidation.

## 9.4   CATALYSTS FOR FOOD OXIDATION

Free radical mediated oxidation has three phases: initiation, propagation, and termination. The initiation phase requires **energy input** to surmount an activation energy barrier. This energy barrier is reduced in the presence of **catalysts**. Common catalysts for food oxidation are listed in Table 9.5.

The main oxidation catalysts are structurally related (Table 9.5, Figure 9.5). First, are the transition metal cations that are well known for their ability to undergo 1-electron reactions. In the heme group, $M^{X+}$ forms a complex with the porphyrin ring. Incorporating the $Fe^{3+}/Fe^{2+}$ cation within a porphyrin ring lowers the redox potential from $770\,mV$ to $-100\,mV$ (water as solvent) or to $+200\,mV$ (nonpolar solvent). The $Fe^{3+} + e \rightleftharpoons Fe^{2+}$ equilibrium favors the $Fe^{3+}$ state in a polar environment where the extra positive charge can be stabilized by hydration, or via electrons donated by porphyrin groups.

Next, the heme group is incorporated in the hydrophobic interior of a globin protein chain. The resulting redox potential ranges from $50\,mV$ (myoglobin) to $-200\,mV$ (peroxidase). In the progression from free iron, heme, globin, and peroxidase the redox characteristics of the $Fe^{3+} + e \rightleftharpoons Fe^{2+}$ half-cell is modified by placing iron inside various environments (Figure 9.5). The functional relations between the different oxidation catalysts are described in the following sections.

### 9.4.1   LIGHT, SENSITIZERS, AND IRRADIATION

Exposing foods to electromagnetic radiation will facilitate oxidation especially if natural dyes and other sensitizers are present (Section 9.3.5). Exposure to particle beam irradiation (gamma rays) will also promote oxidation through the generation of free radical species by homolytic cleavage of covalent bonds.

### 9.4.2   METAL IONS AS OXIDANTS AND PROOXIDANTS

Metal ions promote oxidation by acting as **prooxidants**. In the presence of hydrogen peroxide the transition metal ion help to form (Eq. 9.11) a highly reactive hydroxyl radical. Other **reducing compounds**, by acting as **prooxidants**, can promote oxidation.

**Table 9.5**
**Some important catalysts for food oxidation**

| Catalyst class | Comments |
|---|---|
| Free metal cations ($M^{x+}$) | Transition metal cations, essential minerals |
| Dissociated heme | Porphyrin with a central metal cation, $M^{x+}$ |
| Intact heme proteins | Enzymes (peroxidase, catalase), oxygen binding proteins (hemoglobin, myoglobin), chlorophyll, cytochromes |
| Nonheme metaloproteins | Lipoxygenase, superoxide dismutase |
| Organic reducing agents | Ascorbic acid, glutathione |
| Sensitizers | Dyes, fluorescent compounds |

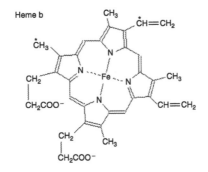

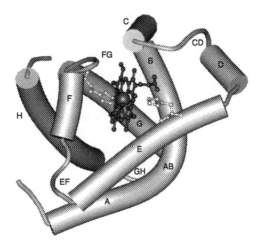

**Figure 9.5** Catalysts for biological oxidation. Top, a dissociated heme molecule; bottom, the heme protein myoglobin.

### 9.4.3 Dissociated Heme

Dissociated heme functions as a prooxidant in the presence of hydrogen peroxide. The process is similar to that described above (Eq. 9.11). Dissociated heme contains a central iron in the $3^+$ oxidation state ($Hm–Fe^{3+}$). This is able to transfer an electron to hydrogen peroxide leading to a heterolytic cleavage of the peroxide molecule to form a hydroxyl ion ($OH^-$) and a heme-bound **hydroxyl radical** (Eq. 9.13) called the **feryl radical** ($Hm–Fe^{4+}–O^•$). The latter species has two resonance structures, $Hm–(Fe^{5+}=O) \leftrightarrow Hm–(Fe^{4+}–O^•)$. Reactions with food molecules (AH) convert these into new radical species ($A^•$) and thereby propagate oxidation. Alternatively, AH can be oxidized in a stepwise fashion to A.

$$Hm.(Fe^{3+}) + HO–OH \rightarrow HOH + Hm.(Fe^{4+}–O^•) \tag{9.14}$$

$$Hm.(Fe^{4+}–O^•) + AH \rightarrow A^• + Hm–(Fe^{4+}–OH) \tag{9.15}$$

$$Hm.(Fe^{4+}–OH) + AH \rightarrow A^• + Hm–(Fe^{3+}) + HOH \tag{9.16a}$$

$$Hm.(Fe^{4+}–OH) + A^• \rightarrow A + Hm–(Fe^{3+}) + HOH \tag{9.16b}$$

The heme group, functioning as a catalyst, is regenerated during the oxidation of AH to A. Oxidation of AH by hydrogen peroxide proceeds more rapidly in the presence of heme than in its absence. The thermodynamics of catalyzed and uncatalyzed reactions are the same. A catalyst provides a different reaction pathway, which leads to a faster reaction.

### 9.4.4  SOURCES OF HEME IN FOODS

The heme group comprises four pyrole rings linked to form a porphyrin ring (Figure 9.5, top)[d]. Nitrogens from the porphyrin ring hold a central metal cation ($Fe^{2+}$, $Mn^{2+}$ or $Cu^{2+}$). Different porphyrin groups differ slightly according to the substituent, R. The heme group (porphyrin ring and central metal ion) is then encircled by a protein coat (apo-protein) to form a range of different heme proteins including **myoglobin**, **hemoglobin**, **peroxidase**, and **cytochrome C oxidase**. Heme proteins function in oxygen transfer as oxidases or as redox proteins. Myoglobin stores oxygen within muscles for release as needed. Catalase and peroxidase are both enzymes that degrade hydrogen peroxide. The characteristics of different heme proteins are due to their different apo-proteins.

Heme binding to the apo-protein coat is usually via hydrophobic interaction. Slight alterations in protein structure during exposure to high temperatures, organic solvents or denaturants, release the heme group into the surrounding medium. Free heme can become attached to the surface of a denatured protein. The denatured heme protein exhibits novel catalytic activity as a consequence of the dissociated heme. In contrast to a heme enzyme such as peroxidase, a free heme group lacks substrate specificity as an oxidation catalyst.[4] Table 9.6 lists the relative efficiency of some different oxidation catalysts against phosphatidic acid liposomes.

## 9.5  ANTIOXIDANTS

### 9.5.1  INHIBITORS OF FOOD OXIDATION

Inhibitors of food oxidation include (1) chelating agents which reduce the concentration of free metal cations in solution, (2) free radical quenchers which form stable and less reactive free

**Table 9.6**
**The relative efficiency of different oxidation catalysts against phosphatidic acid liposomes**

| Compound | Rate ($n$ mole $O_2$ consumed per min per $n$ mole heme) | Rate of reaction (% maximum) |
|---|---|---|
| Hematin | 348 | 100 |
| Microperoxidase | 320 | 94 |
| Met. myoglobin | 19 | 5.5 |
| Horseradish peroxidase | 1.6 | 0.5 |
| Cytochrome P450 | 0.2 | <0.01 |
| Protoporphyrin | 0 | 0 |

[d] In addition to the term "heme" the reader may encounter the terms hemin and hematin. Heme is a porphyrin ring containing iron in the oxidation state $2^+$ (ferrous iron). Hemin contains $Fe^{3+}$ complexed with chloride. Hematin is a heme group with a $Fe^{3+}$ central cation complexed with the hydroxyl ion as counterspecies.

**Table 9.7**
**Classes of antioxidants**

| Antioxidant | Mode of action |
|---|---|
| **Synthetic** | |
| Butylated hydroxyanisole (BHA) | 2 |
| Butylated hydroxytoluene (BHT) | 2 |
| Tertiary-butylated hydroxyquinone (TBHQ) | 2 |
| Propyl gallate | 2 |
| Ethylenediamine tetraacetic acid (EDTA) | 1 |
| **Natural** | |
| Ascorbic acid (vitamin C) | 3 |
| Carnosine (and amino acids and peptides (cysteine, histidine, glutathione)) | |
| Flavonoids | 1–3 |
| L-ascorbyl palmitate | 3 |
| Phytic acid | 1, 3 |
| Polyphenols | 1, 2 |
| Tocopherol (vitamin E) | 2 |
| Thiol compounds (e.g., glutathione) | 3 |
| Thiol proteins | 3 |
| Maillard reaction products | 2, 3 |
| Enzymes (superoxide dismutase, catalase) | |
| **Plant extracts** | |
| Rosemary | 1–3 |
| Onion | 1–3 |
| Ground mustard | 1–3 |
| Tea extract | 1–3 |

Mode of action: 1, chelation; 2, free radical quencher; 3, reducing agent.

radicals, (3) reducing agents (electron donors), and (4) agents that remove or reduce the concentrations of oxygen and/or hydrogen peroxide. Food antioxidants can also be classed as **synthetic** or **natural** (Table 9.7). Antioxidants are added to foods during preparation and storage.[5] Supplementing livestock feed with high levels of an antioxidant (vitamin E) protects meat from oxidative damage.

## 9.5.2 PHYSICAL PREVENTION OF OXIDATION

Oxidation can be prevented by physical factors. Finely divided foods have a higher surface area for gaseous exchange and oxidation. Grinding and shearing can cause the release of oxidation catalysts. Refrigeration is not an efficient method for preventing oxidation because the solubility of oxygen increases at low temperatures. Physical methods for preventing oxidation include the use of packaging, and packaging materials have been developed that contain active (reducing) agents. Another important innovation is **vacuum packaging** whereby the food container is evacuated to remove oxygen. Vacuum packaging is especially effective with meat products.

## 9.5.3 MEASUREMENT OF TOTAL ANTIOXIDANT CAPACITY

There is increasing interest in antioxidants owing to their health promoting action as well as for their ability to prevent lipid oxidation in processed foods. **Dietary antioxidants** appear to promote the body's resistance towards free radical mediated disease states including premature aging.[6] Methods for determining **total antioxidant capacity** of food extracts include:

(1) Evaluation of antioxidant action using a model reaction based on the oxidation of linoeleic acid or human low-density lipoprotein (LDL) *in vitro*.
(2) Assessment of radical trapping power. Methods include the DPPH (2,2-diphenyl-1-picrylhydrazine) radical assay, the TRAP (total radical-trapping parameter) assay, the TEAC (Trolox equivalent antioxidant activity) assay, and the ORAC (oxygen radical absorbance capacity) assay.
(3) FRAP (ferric-reducing antioxidant power).[7]

The **TEAC assay** follows antioxidant *inhibition* of **crocin**[e] bleaching by the peroxyl (superoxide) radical. The results are expressed as the weight of $\alpha$-tocopherol able to produce the same anti-bleaching action as the test sample. Results for water soluble antioxidants are expressed in terms of the Trolox equivalent antioxidant activity (TEAC). The TEAC value is the weight of **Trolox C** (Figure 9.6) that produces the same anti-bleaching effect as the test sample. The TEAC assay has quickly become the "gold standard" for measuring dietary antioxidants and will be described briefly below.

The key components of the TEAC assay are two azide compounds, ABAP (2,2'-azo-bis(2-amidinoprop'ane)) and AMVN (2,2'-azo-bis(2,4-dimethylvaleronitrile)), which are used for developing tests for oil soluble and water soluble antioxidants, respectively. At high temperatures azide compounds (cf. general formula R–N=N–R for an azide) dissociate to form radical species R*. This then reacts with diatomic oxygen to form the radical ROO*. In the presence of carotenoid (crocin) the formation of ROO* results in a bleaching action and a loss of color at 443 nm. The net reaction taking place in the absence of added antioxidant is summarized in Eq. 9.20. Furthermore, Eqs. 9.20a–c show some important steps for the free radical bleaching of carotenoid dyes:

$$R-N=N-R + 2O_2 + 2(crocin) \rightarrow 2ROOH + N_2 + 2(crocin^*), \tag{9.17}$$

$$R-N=N-R \xrightarrow{heat} 2R^* + N_2, \tag{9.17a}$$

$$2R^* + 2O_2 \rightarrow 2ROO^*, \tag{9.17b}$$

$$2ROO^* + 2(crocin) \rightarrow 2ROOH + 2(crocin^*). \tag{9.17c}$$

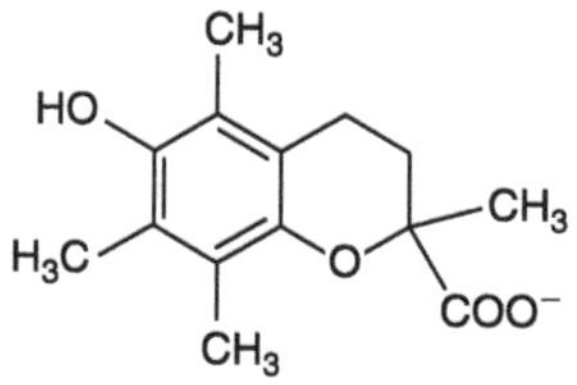

**Figure 9.6**  Trolox C, a water-soluble analogue for vitamin E ($\alpha$-tocopherol).

---

[e]Crocin is the coloring matter of Chinese yellow pods, the fruit of *Gardenia grandiflora*. It is a water soluble carotenoid with a structure resembling $\beta$-carotene. As a red powder made from saffron (which is obtained from *Crocus sativus*) crocin is also called saffronin or polychroite. The coloring matter of saffron changes color upon treatment with certain acids. Crocin can be readily prepared by extracting saffron with methanol. This natural product should not be confused with a brand of high dose paracetamol (Crocin-1000) manufactured by GlaxoSmithKline Asia Private Limited.

With antioxidant present, the rate of crocin bleaching slows owing to the extra reaction step (Eq. 9.18) though the final products are the same as above.

$$ROO^* + antiOx \rightarrow antiOx^* + ROOH \tag{9.18}$$

During the TEAC assay, the initial rate of crocin bleaching is measured without ($v_0$) and with antioxidant present ($v_A$). A graph of $v_A/v_0$ is then plotted versus [antioxidant]/[crocin]. The gradient of this graph equals $k_A/k_0$ (where $k$ is a rate constant and the subscripts have their previous meaning). The TEAC value is determined by dividing $k_A/k_0$ with a similar ratio determined for Trolox C.

## 9.6 PRACTICAL ASPECTS OF LIPID OXIDATION

### 9.6.1 MEASUREMENT OF PEROXIDE VALUE

Hydroperoxides produced by lipid oxidation reduce iodide to iodine at low pH. The resulting brown color can be measured by colorimetry. **Peroxide value** (PV) is the concentration of hydroperoxide measured as milliequivalents of iodide reduced per kg of oil (meq/kg). The reaction between hydroperoxide and iodide leads to the eventual formation of the corresponding alcohol (Eqs. 9.14a and 9.19b):

$$2I^- \rightarrow +I_2 + 2e \ (E^\circ = -0.54V) \tag{9.19a}$$
$$RCOO + 2e + 2H^+ \rightarrow H_2O + RCOH \tag{9.19b}$$

### 9.6.2 THIOBARBITURIC ACID TEST

The thiobarbituric acid (TBA) test for lipid oxidation measures **molanaldehyde**, which is a 3-carbon dialdehyde secondary product of lipid oxidation. Molanaldehyde reacts with TBA forming a red pigment having an absorbance maximum at 532 nm (Figure 9.7). Test results are normally given as the **TBA number**, which is the concentration of molanaldehyde in mg per kg oil. The TBA test is performed on whole food, extracts or a steam distillate. Analysis using a food distillate produces results that are 1.5–2 times higher than results from whole foods. As well as molanaldehyde, TBA also reacts with 4-hydroxyalkenyl (4-HAD), another product of lipid oxidation. The TBA method is criticized for its possible lack of specificity. Molanaldehyde can also be degraded after prolonged food storage. Despite such disadvantages the technique is widely applied. Figure 9.7 shows the reaction between two molecules of 2-thiobarbituric acid and molanaldehyde.

### 9.6.3 OTHER DETECTION METHODS FOR LIPID OXIDATION

Carbonyl compounds formed during lipid oxidation can be determined with **2,4-dinitrophenylhydrazine**. Diphenylhydrazone products formed with carbonyls absorb light at 340 nm. Other important methods for monitoring lipid oxidation include the determination of volatile products via headspace **gas–liquid chromatography**. Alternatively, the fall in oxygen concentration can be monitored using the **Clark oxygen electrode**. For non-turbid samples lipid oxidation can be followed by determining the increase in **UV absorbance** at 233 nm, as a polyunsaturated fatty acid (usually linoleic acid) is oxidized and converted to a conjugated fatty acid hydroperoxide.

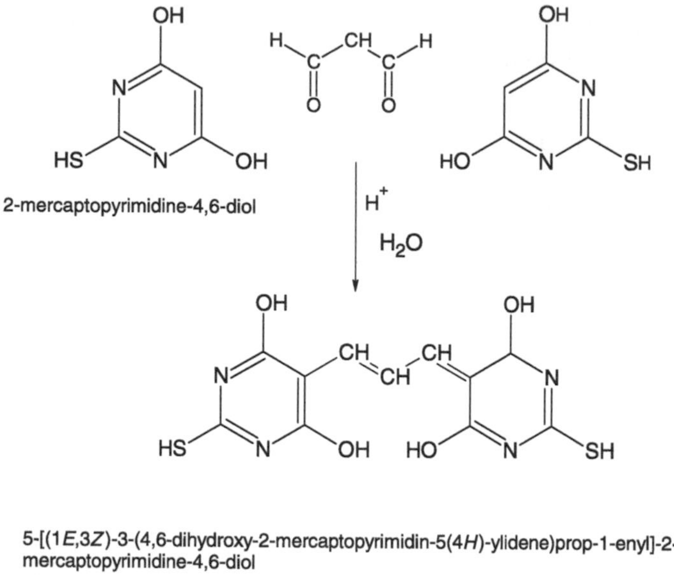

2-mercaptopyrimidine-4,6-diol

$H^+$

$H_2O$

5-[(1$E$,3$Z$)-3-(4,6-dihydroxy-2-mercaptopyrimidin-5(4$H$)-ylidene)prop-1-enyl]-2-mercaptopyrimidine-4,6-diol

**Figure 9.7** Two molecules of 2-thiobarbituric acid (i.e., 2-mercatopyrimidine-4,6-diol) react with molanaldehyde to form a red TBA–molanaldehyde pigment.

### 9.6.4 WARM-OVER FLAVOR IN MUSCLE FOODS

Some consequences of lipid oxidation were summarized in Figure 9.1. A well-known cause of quality loss for **cooked meat** products and pre-cooked meals is **warm-over flavor** (WOF). This is a loss of fresh meat aroma and its replacement by a "cardboard" aroma within 1–3 days of refrigerated storage.[8,9] According to one hypothesis WOF is the result of lipid oxidation catalysed by free iron ($Fe^{2+}$) or non-heme iron. Alternatively, lipid oxidation is catalysed by heme probably derived from myoglobin, hemoglobin and/or cytochrome.

WOF was studied using samples of ground beef thoroughly washed to remove endogenous inhibitors or catalysts of lipid oxidation. Then, known amounts of free iron (also called non-heme iron), myoglobin (a source of heme groups), and/or lipid was added. After several days in a fridge, samples were assayed for oxidation using the TBA method. The investigations showed that myoglobin (1–10 mg/mL) did not promote WOF and that phospholipids were better substrates for WOF formation as compared to normal triglycerides. $Fe^{2+}$ was more important as a catalyst for WOF. These findings were later questioned as unrepresentative of real meats. Washing ground beef was thought to remove traces of hydrogen peroxide, which may be required for oxidation. When hydrogen peroxide was added to the test system myoglobin was found to be the more efficient catalyst of lipid oxidation compared to the ferrous ion (Figure 9.8). Table 9.6 also provides some idea of the relative importance of various catalysts for lipid oxidation.

Meat samples will show varying susceptibility to WOF due to differences in their concentrations of PUFA, iron, vitamin E (tocopherol), and other antioxidants. Red poultry meat is more susceptible to WOF as compared to white owing to the higher content of iron and myoglobin in the former. The incidence of WOF can be controlled using (1) synthetic antioxidants including nitrite, phosphate, butylated hydroxyanisole (BHA), butylated hydroxy toluene (BHT), $t$-butylated hydroxyquinone (TBHQ), and propyl gallate; (2) vitamin E, ascorbic acid, and other natural antioxidants; and (3) pre-slaughter **dietary supplementation**. Adding antioxidants to meat can extend its storage life by reducing oxidative damage.

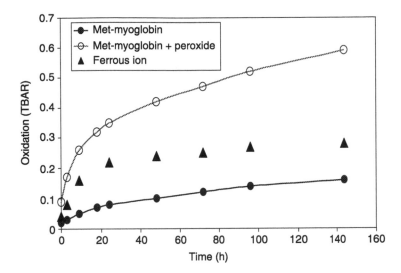

**Figure 9.8** The efficiency of met-myoglobin and ferrous ion ($Fe^{2+}$) as catalysts for lipid oxidation in model systems comprising washed minced beef. Oxidation was measured as the thiobarbituric acid (TBA) value, which is nanomoles of molanaldehyde formed per mg of protein. (Drawn using data from Asghar et al.[8])

Addition of tocopherol to the diets of pigs, broiler chickens, and deer produced up to 77% reduction in TBA number of stored meats.[5]

Lipid oxidation in seafood is another area of great concern. Up to 60% of mackerel products can be rendered unfit for human consumption due to oxidation. Oxidation follows a lag period as expected for a free radical reaction. The rate of oxidation is related to (1) the nature of fatty acids in the fish, (2) distribution of TAGs and phospholipids in the body, (3) chemical factors including the presence of accelerators (metal ions), antioxidants, and pH, and (4) external factors, such as temperature and light. Storage studies with sardine or mackerel show that the rate of deterioration of skin lipids is 4–5 times higher than those of muscle lipids probably because of exposure to atmospheric oxygen. Glazing was an effective method for lowering the rate of oxidation of stored fish.[10]

### 9.6.5 OFF-FLAVORS DURING MALTING AND BREWING

A loss of flavor occurs during malting and brewing due to oxidation. The flavor defects are due to aldehydes and carbonyl compounds formed by oxidation of PUFA. Barley malt contains a host of components that affect oxidation including endogenous enzymes (catalase, peroxidase, polyphenol oxidase), antioxidants (ascorbic acid, glutathione, polyphenols). Flavor deterioration was also affected by increased oxygen exposure caused by turbulence and agitation during the transfer of wort. The oxygen content of ingredients and level of free metal ions (e.g., copper ions) also affect the flavor stability of beer.[11]

### References

1. Richardson, T. and Korycka-Dahl, M., Lipid oxidation, in *Developments in Dairy Chemistry*—2, Fox, P. F., Ed., Applied Science Publishers, New York, 241–363, 1983.
2. Wong, D. M. S., *Mechanism and Theory in Food Chemistry*, Van Nostrand Reinhold, New York, 401–403, 1989.

3. Segel, I. H., *Biochemical Calculations*, 2nd edition, J. Wiley & Sons, New York, 172, 1976.
4. Kaschnitz, R. M. and Hatefi, Y., Lipid oxidation in biological membranes. Electron transfer proteins as initiators of lipid auto oxidation, *Arch. Biochem. Biophys.* 171, 292–304, 1975.
5. Mielche, M. M. and Bertelsen, G., Approaches to prevention of warmed-over flavor, *Trends Food Sci.* 5, 322–327, 1994.
6. Halliwell, B. and Gutteridge, M. C., Role of free radicals and catalytic metal ions in human disease: An overview, *Methods Enzymol.* 186, 1–85, 1990.
7. Frankel, E. N. and Meyer, A. A., The problems of using one-dimensional methods to evaluate multifunctional food and biological antioxidants, *J. Sci. Food Agric.* 80(13), 1925–1941, 2000.
8. Asghar, A., Gray, I. J. U., Buckley, D. J., Person, A. M. and Booren, A. M., Perspectives in warmed-over flavor, *Food Technol.*, June, 102–108, 1988.
9. Ladikos, D. and Lougvois, V., Lipid oxidation in muscle foods: a review, *Food Chem.* 34(4), 295–314, 1990.
10. Khayat, A. and Schwall, D., Lipid oxidation in sea food, *Food Technol.*, July, 130–140, 1983.
11. Bamforth, C. E., Muller, R. E. and Walker, M. D., Oxygen and oxygen radicals in malting and brewing. A review, *Am. Assoc. of Brewing Chem. J.* 51(3), 79–88, 1993.

# 10

# The Maillard Reaction

## 10.1 INTRODUCTION

Three types of browning reactions occur in foods. First is the **Maillard reaction**, which results from a reaction between carbonyl compounds (aldehydes, ketones, and reducing sugars) and amines. Second, there is **caramelization**, which occurs when sugars are heated to high temperatures. *Third*, there is **enzymatic browning** catalyzed by enzymes such as polyphenol oxidase, lipoxygenase, and peroxidase.[1]

Nonenzymic browning (NEB) begins with a simple addition reaction between a reducing sugar and a primary amine. The *end* product is a brown pigment called **melanoidin**. The middle part of the Maillard reaction is not wholly understood despite 50 years or more of intensive study. The path for NEB changes with temperature, time of reaction, composition of the food, presence of moisture, and pH.[2-5] Parallel or sequential reactions produce numerous **Maillard reaction products** (MRPs) most of which are only now being identified. In foods the amine components for NEB are amino acids, polypeptides or proteins. A range of carbonyls formed from reducing sugars, lipids, and ascorbic acid undergo NEB. The importance of MRPs is due to their role as **flavors**, **colors**, and **antioxidants**. In some foodstuffs, most notably dried milk powder, NEB leads to unwanted discoloration, and loss of protein nutritional value. The color and aroma of baked products (e.g., bread and cookies) and roasted meat is partly the result of MRPs. Roasted coffee contains high concentrations of MRPs.

MRPs have a number of adverse characteristics. Some early MRPs formed by reacting sugars and lysine include glycated proteins having reduced digestibility. High temperature cooking such as grilling or frying leads to **heterocyclic amines** that are potent **mutagens**. There was great furor recently when Swedish scientists reported that **acrylamide**, a known carcinogen, is produced during the Maillard reaction.

The reaction between sugars and amino acids occurs if such **ingredients** are heated together in solution. The Maillard reaction also occurs during prolonged storage of foodstuffs at room temperature. Digestive enzymes cannot break down covalent bonds formed by MRPs. Protein glycation reduces their **nutritional value**. Chronic high blood glucose levels produce **glycation** of serum proteins. The degree of glycation is used to diagnose the potential for diabetes long before the full-blown disease occurs. The Maillard reaction also leads to the formation of **advanced glycation end-products** (AGEs). Long-lived proteins in the body, including tendon collagen and crystalline, a globular protein found in the lens of the eye, form fluorescence or yellow deposits. Glycation of eye-lens protein is implicated in the formation of eye cataract. The formation of AGEs has received a great deal of attention from biomedical scientists who suspect that this is involved in a great many degenerative diseases, including arthritis and Alzheimer's disease. Some effects of the Maillard reaction are listed in Table 10.1.

---

**Table 10.1**
**Some effects of Maillard reaction in foods**

- Formation of pleasant food flavors and odors
- Development of desirable colors and tones
- Formation of browning reaction products as part of caramelization
- Formation of antioxidants
- Formation of reactive oxygen species
- Destruction of protein nutrient value
- Formation of mutagenic products
- Formation of advanced glycation end-products

---

Clearly, the Maillard reaction pathway leads to "good and bad" effects in foods. Much has been written about the generation of flavors and colors via NEB. In this chapter, the focus is to consider the Maillard reaction and its role in food deterioration. Following a general introduction, Section 10.2 provides a brief summary of the chemistry of the Maillard reaction. Section 10.3 considers the MRPs which function as colors and flavors, potential mutagens or antioxidants. Section 10.4 explains how conditions such as moisture content, temperature, extrusion cooking, and pH affect the rate of the Maillard reaction. In Section 10.5 we examine the Maillard reaction and its link with nonenzymatic oxidation (Chapter 9).

## 10.2   CHEMISTRY OF THE MAILLARD REACTION

The accepted scheme for NEB was first described by Hodge (1953).[1] The reactions begin with an addition–elimination step involving an aldehyde and a primary amine. Carbohydrates are the commonest aldehydes. Simple sugars are converted to *dicarbonyl compounds* with the regeneration of a catalytic primary amine. The highly reactive dicarbonyls then react with themselves, amines, amino acids etc. to form MRPs.

Vitamin C is another important browning substrate. Ascorbic acid undergoes browning following oxidation to dehydroascorbic acid and hydrolysis to a dicarbonyl compound **2,3-diketogluconate (2,3 DKG)**. The conversion of ascorbic acid to DKG is the normal route for vitamin C destruction in foods. Like other dicarbonyls, DKG reacts with amines to form brown pigments. A further group of dicarbonyl compounds involved in browning include natural cell constituents such as **dihydroxyacetone** and **methylglyoxal**. Frequently, the primary amine ($ANH_2$) for NEB comes from amino acids, the N-terminal amine of peptides and proteins, and from protein amino acid side chains of lysine and arginine. Sources of dicarbonyl compounds taking part in the Maillard reaction are shown in Figure 10.1 and a summary of the reactions involving sugars is shown in Figure 10.2. The reactions leading to MRPs are summarized below:[6,7]

(1) **Formation of Schiff's base**. A primary amine $ANH_2$ reacts with a sugar $R(CHOH)_4CHO$ via addition–elimination. Nucleophilic attack at the carbonyl group is followed by the loss of water to form an imine or Schiff's base (**I**).

(2) **Amadori rearrangement**. The imine (**I**) (Figure 10.1(b)) isomerizes to form the Amadori compound and the C1=N double bond migrates to C2=C1 (**IIa**). At pH < 4, the Amadori rearrangement results in a further double migration to C3=C2 so forming the 2,3 enol compound (**IIb**). Both *Amadori intermediates* (**II**) undergo **enol–keto isomerism** to form $\alpha$-aminoketones (not shown). The **Heynes rearrangement** is the analogous reaction for fructose and other ketoses.

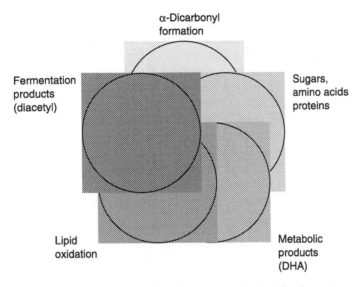

**Figure 10.1**  Sources of dicarbonyl compounds taking part in the Maillard reaction.

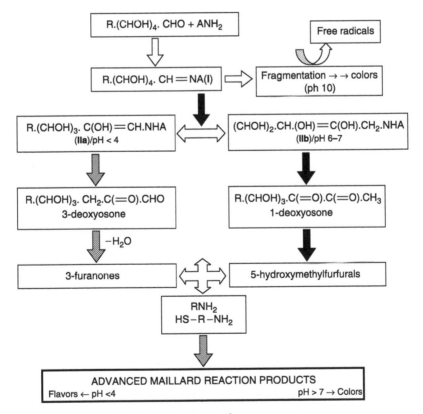

**Figure 10.2**  Summary of early Maillard reaction involving sugars.

(3) **Formation of deoxyosones**. The Amadori intermediate (**II**) releases an amine (ANH$_2$) and forms deoxyosone in a pH-dependent fashion. At low pH (< 4) the 1,2 enol (**IIa**) forms **3-deoxyosone**. At intermediate pH (5–7) compound **IIb** decomposes to form the **1-deoxyosone**.

(4) **Flavor compound formation from deoxyosone**. Reactive **dicarbonyl** deoxyosones undergo further reactions leading to flavor and color compounds. For instance 1-deoxyosone cyclizes/dehydrates so forming 5-methylfurfural and 5-hydroxy-methylfurfural at pH 6–7. In contrast, **3-deoxyosone** forms 3-furanone at pH < 4. The condensation of these ring structures is the first step towards fluorescence and then colored Maillard reaction products, and ultimately the melanoidins.

(5) **Degradation of Amadori products at high pH.** Under alkaline conditions, the Amadori product fragments to form small molecular weight dicarbonyl compounds which self-polymerize to form melanoidins.

(6) **Strecker degradation**. Some 2- and 3-carbon dicarbonyl compounds combine with amino acids and subsequently eliminate RCHO leading to simple heterocyclic ring compounds including pyrazines, oxazoles, thioloze, pyrrole, and thiophenes.

Hodge divided the Maillard reaction into three major steps.[1] The initial stage leads to colorless compounds with zero absorption for UV light. The major reaction mechanisms are amine condensation and the Amadori rearrangement. The intermediate stage of NEB forms colorless or yellow products with strong absorption in the UV region. Intermediate products arise from sugar dehydration, fragmentation, and amino acid degradation. The final stage of NEB, leading to highly colored products, involves aldol condensation and aldehyde–amine polymerization reactions. Hodge's account, with minor alterations and embellishments over the past 50 years, remains accurate to this day.

## 10.3   PRODUCTS OF THE MAILLARD REACTION

### 10.3.1   MAILLARD REACTION COLORS

Early MRPs have low molecular weights and absorbancies for UV light. The next sequence of products is **fluorescent** compounds, followed by yellow, orange, red, wine red, and brown products. The succession of Maillard reaction *colors* results in a mixture of **pigments** and **dyes**. A pigment is a light-absorbing species, which, owing to its large molecular weight, also scatters light. The colors are the subset of MRPs with absorbance in the visible wavelength range.

As NEB proceeds beyond the deoxyosone stage (Figure 10.2) there is increasing polymerization. The extent of double bond conjugation also increases and passes through a maximum with increasing color formation. However, as polymerization proceeds the net amount of unsaturation must decline. The absorbance spectra for Maillard reaction colors are usually broad starting from the UV and extending to the visible wavelength range. There is no single maximum wavelength for absorption ($\lambda_{max}$) because the final spectrum is a mathematical summation of spectra from many different chemical species. For convenience, the formation of the brown color is generally measured at 420 or 500 nm.[8]

**Low molecular weight** Maillard reaction colors have been identified by reacting sugars with ammonium compounds such as isopropyl ammonium acetate and then extracting with organic solvents followed by extensive characterization. These studies found products with a molecular weight of < 500 Da and an absorbance peak at about 360 nm. The earliest colors form by the condensation of furfural and/or furanones produced via the cyclization of deoxyosone (Figure 10.2). When pentoses were heated with alanine in a ratio of 10:1 in an aqueous solution at pH 7.0 a yellow product identified as 2-[(2-furyl) methylidene]-4-hydroxy-5-methyl-2*H*-furan-3-one was formed.[9]

The **high molecular weight** colors have not been characterized to any great extent. NEB pigments or melanoidins are formed after prolonged heating of sugars and amines. Polymeric colors also result when starch and proteins react with amino acids and sugars, respectively, forming **pendant** Maillard reaction colors. In these structures, the dye molecules are attached to a biopolymer backbone. Pendant Maillard reaction colors can be released following enzymatic digestion by amylases and/or proteases, suggesting that protein side chains can undergo the Maillard reaction.[8–11]

## 10.3.2 MUTAGENIC MAILLARD PRODUCTS

### 10.3.2.1 Diet Related Cancer

The toxicological effects of MRPs have received much attention recently.[12] It was suggested that some MRPs may be linked with cancer. The World Cancer Research Foundation (WCRF) and American Institute of Cancer Research (AICR) report published in 1997 noted that 30–40% of cancers are food related and could be prevented by modifying the diet.[13] The 670-page report (with 4500 references) reviewed epidemiological evidence linking dietary patterns with the incidence of different types of cancers. Tobacco related cancer (affecting the mouth, neck, larynx, and/or lungs) was foremost throughout the world. Smoking as well as chewing tobaco could lead to cancer.

Diet related cancers (affecting the stomach, colon, and rectum) were the second most common forms of cancer. They were associated with high fat, high salt, and low fiber diets in developed countries. The highest rates of stomach cancer were found in Japan where it is the principal cause of cancer deaths, linked to the consumption of cured, smoked, salted, pickled or heavily spiced meat. In the United States colorectal cancer (cancer of the colon and rectum) was the major diet related cancer, accounting for 1.2 million cases and about 60,000 deaths annually.[14] Furthermore, the incidence of colorectal cancer in the US increased with age: the yearly death rate for 15–34, 35–54, 54–74, and 74+ year-olds was 500 persons, 5200 persons, 28,400 persons or 26,000 persons, respectively.[14]

Consumption of thermally processed **high-protein foods** was a contributary factor for diet related cancer. In contrast, vegetable and high fiber diets decreased cancer incidence.[13] The WCRF/AICR report concluded that

*Cancer is principally caused by environmental factors, of which the most important are tobacco; diet and factors related to diet, including body mass and physical activity; and exposures in the workplace and elsewhere. These [factors] interact with the differing vulnerability, both inherited and acquired, of individuals' constitutions. Much of the world's cancer burden could therefore be prevented if people did not smoke tobacco, by appropriate dietary and activity patterns, and by reducing other environmental exposures.*[13]

The level of mutagens within the general diet is a matter of continued interest.

### 10.3.2.2 Dietary Mutagens from Thermally Processed Foods

Dietary mutagens[a] form during the Maillard reaction at high temperatures.[15] For instance, **dicarbonyl compounds** occur in roasted coffee whilst **acrylamide** is found in some snack foods including potato chips.[16] **Heterocyclic amines** (HCAs), or heterocyclic aromatic amines (HAAs), which are potent mutagens, were detected in grilled meat. Structures of some heterocyclic amines are shown in Figure 10.3.

---

[a]A mutagen is a compound that, according to the result of the Ames test or DNA damage test, produces mutations in bacteria. Carcinogens are compounds which, when administered to test animals, produce tumors and cancers. A strong correlation exists between mutagenic activity and the ability to cause cancer *in vivo*. However, not all mutagens are carcinogens. An antimutagen is any agent that prevents DNA damage by mutagens.

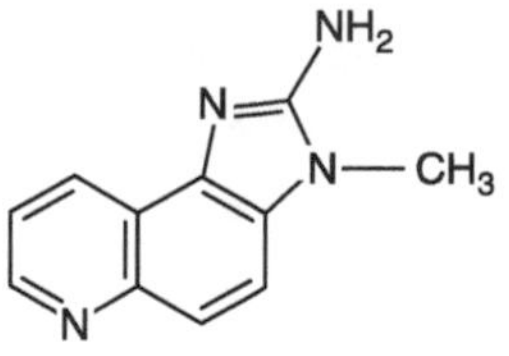

| IQ | MelQ |
|---|---|
| 2-Amino-3-methylimidazo-[4, 5-7] quinoline | 2-Amino-3, 4-0 methylimidazo-[4, 5-7] quinoline |

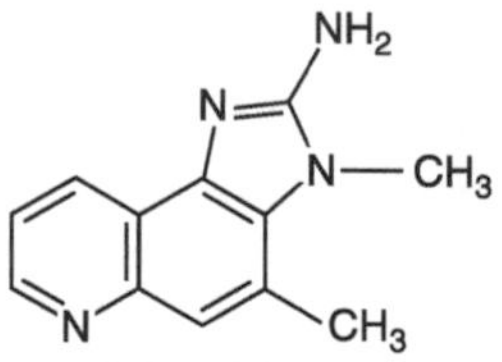

**Figure 10.3**  Structures of some heterocyclic amines.

---

**Table 10.2**
**Maillard reaction products with mutagenic activity**

| Maillard reaction products | Comments | Mutagenicity* |
|---|---|---|
| Dicarbonyl compounds | | |
|   Glyoxal | Coffee, baked bread crust | Yes |
|   Methylglyoxal | Naturally present in cells | Uncertain in humans |
|   Diacetyl | Buttery flavor, popcorn flavor | |
| Furans | Caramel-like odor, abundant MRP | Yes, safe at low concentrations |
| Pyrroles | Pleasant flavor, corn-like odor | None reported, in absence of nitrite |
| Diathianes (1,3 or 1,4) | | Yes |
| Thiazolidines | Meaty flavor | Yes |
| Imidazoles | | |
| Pyrazines | Most common heterocyclic formed via the Mailliard reaction | None, possible antimutagen |
| Acrylamide | Formed by heating carbohydrate-rich foods | Neurotoxin, carcinogen |

Yes = compound shows mutagenic activity though "weak" by comparison to HCA

---

Acrylamide and HCA are **promutagens**, which need to undergo oxidation by cytochrome P450 to become harmful. According to a prevailing model, **food mutagens** are electrophiles, which react with electron-dense (nitrogen) sites on the DNA chain to form addition products. DNA-adduct formation is an essential step for mutagenesis.[17]

A common assay for mutagens uses the Ames test.[18–20] This employs special strains of mutant *Salmonella typhimurium*, which have a defective metabolic pathway for synthesizing the amino acid histidine. Mutant *S. typhimurium* fail to grow unless histidine is added to their growth media. Alternatively, the bacteria can undergo spontaneous mutations in their DNA and thereby attain the capacity to manufacture histidine. In the presence of a mutagen, the *frequency* of DNA mutations is greatly increased leading to the restoration of the histidine biosynthetic pathway. The resulting bacteria grow in histidine-free media. In short we can evaluate different mutagens by their ability to encourage the growth of mutant *S. typhimurium* in a medium lacking histidine.

Several MRPs appear to act as mutagens (Table 10.2). The best understood are probably dicarbonyl compounds formed when deoxyosones fragment to low molecular weight products (Table 10.3). Mutagenicity tests using *S. typhimurium* TA100 or T98 show the order of mutagenicity is glyoxal > methyl glyoxal > phenyl glyoxal $\gg$ 1,2-cyclohexanedione $\gg$ diacetyl > 3,4-cyclohexanedione. Mutagenicity correlates with the reactivity of dicarbonyl compounds with purine nucleotides[21]. Studies with 5-hydromethyl furfural (5-HMF) indicate that this compound is probably nonmutagenic at the concentrations used as a flavor.

**Table 10.3**
**Structures of some dicarbonyl compounds**

$$\begin{array}{cc} O & O \\ \| & \| \\ HC & -CH \end{array}$$

| | | |
|---|---|---|
| • H | • H | Glyoxal |
| • –CH₃ | • H | Methylglyoxal |
| • –CH₃ | • –CH₃ | Diacetyl |
| • H | • Phenyl | Phenylglyoxal |

The pyrrole MRPs were nonmutagenic when tested with *S. typhimurium*. Nitro-pyrrole derivatives, formed in nitrite containing foods, may be mutagenic.

HCAs are produced when meat is cooked at temperatures of about 300°C. The imidazole [4,5-*f*] quinoline (IQ) ring structure (Figure 10.3) forms when a mixture of creatine/creatinine, sugars, and amino acids are heated to high temperatures. Pyrolysis of tryptophan at high temperatures is believed to lead to a second group of HCAs having a PhIP (2-amino-1-methyl-6-phenylimidoso[4,5-b]pyridine) ring.[22] It has also been suggested that PhIP is formed due to the condensation of phenylalanine and creatinine.[23] Some HCAs were found to be 550–640 times more harmful compared to benzo[*a*]pyrene, which is a known mutagen.[12]

The likely connection between stomach cancer and the consumption of cooked meat was described above. According to a National Cancer Institute (1996) report:

*Those who ate their beef medium-well or well-done had more than three times the risk of stomach cancer than those who ate their beef rare or medium-rare. People who ate beef four or more times a week had more than twice the risk of stomach cancer than those consuming beef less frequently. Additional studies have shown an increased risk of developing colorectal, pancreatic, and breast cancer associated with high intakes of well-done, fried, or barbequed meat.[24]*

Variables that affect HCA formation include:

(1) **Cooking temperature**. There was 2–3-fold increase in the formation of heterocyclic amines as cooking temperatures were increased from 200°C to 250°C.
(2) **Cooking method**. Frying, broiling, and barbecuing, which involve the highest cooking temperatures, produce the largest quantities of heterocyclic amines. Oven baking also leads to HCAs. These mutagens are also found in the dripping from roasted meat. In contrast to oven cooking most other forms of cooking using water or steam (broiling, poaching) do not appear to produce HCAs. The same is true for microwave cooking, where temperatures do not exceed 100°C.
(3) **Type of food**. The main sources of HCAs were found in cooked muscle meats whereas low HCA levels were associated with milk, eggs, tofu, and meat offal.
(4) **Cooking time**. The formation of HCAs increased with cooking time. The shorter cooking times applied to fast foods led to fewer heterocyclic amines compared with normal domestic cooking. For reviews on this subject see Stavric[25] and Adamson and Thorgeirsson.[26]

### 10.3.3 ANTIOXIDANT MAILLARD PRODUCTS

It was reported in the 1950s that preheating fluid milk before spray drying improves storage stability towards oxidation. There are at least two explanations for this effect. First, heating milk exposes protein sulfhydryl groups, which are masked in unheated milk. These sulfhydryl groups are short-lived when exposed to air and oxidize rapidly. The second explanation for the increase in oxidation stability of preheated milk is increased formation of furfural, which functions as an antioxidant.The concentration of **furfural** increases due to reactions between lactose and milk polypeptides.

A direct link between antioxidant formation and NEB is confirmed by more recent studies.[27] The new experiments employed model systems involving well-defined mixtures of sugars and amino acids. Typically, a mixture of glucose, fructose or sucrose and one amino acid was heated at 95–100°C for 5–50 h. The extent of NEB was measured from absorbance changes at 420–480 nm. Antioxidant capacity was measured by one of the methods described in Section 10.5.3. Variables found to affect the rate of antioxidant and color formation include heating temperature and time, initial pH, type of amino acid, and the molecular weight of any MRPs formed. In many natural foods, including pasta and **tomato puree**, antioxidant capacity develops with the appearance of fluorescent or colored MRPs. Positive correlations were established between browning extent and antioxidant formation. Apparently, confounding results were obtained for those foods with high initial levels of natural antioxidants, which initially declined during the Maillard reaction.

## 10.4  VARIABLES AFFECTING THE MAILLARD REACTION

### 10.4.1  WATER AND TEMPERATURE

A range of storage conditions and processing variables, including temperature–time, moisture, pH, and the composition of the system, affect the Maillard reaction. The effect of water activity ($A_w$) on food deteriorative processes was widely studied during the 1970s. Before discussing the detailed results, we shall review some fundamental characteristics of water as a food component. It is well known that electrons between oxygen and hydrogen are not equally shared owing to the former element being more electronegative. This results in the water molecule being polarized. The strong dipole–dipole interactions lead to **hydrogen bonding** (Figure 1.7) and an unexpectedly high boiling point of liquid water. Interactions between polar water molecules and polar solutes, salts and sugars, results in their dissolution. Hydration interactions also account for the ionization of acids and bases. Finally, water–solute interactions lead to **colligative properties** such as the **elevation of boiling point**, **depression of freezing** point, and changes in **osmotic pressure**.

Moisture uptake by dry or semi-dry (so-called intermediate moisture) foods affect their **shelf-life** and **stability**. Moist products succumb to bacterial spoilage more quickly than thoroughly dried foods. The prolonged shelf-life of dried foods is due to the inhibition of chemical, enzymatic and microbial deterioration processes at low relative humidity. The activity of most enzymes ceases below a relative humidity of 50%. Microorganisms cease to grow below a relative humidity of about 70% (Figure 10.4). Moisture–stability relations provide useful guidelines for food storage.[28]

The rate of NEB peaks at $A_w = 0.6$–0.7 then decreases as explained by one of the following models:

(1) **Diffusion model**. NEB decrease at very low $A_w$ due to diffusion restrictions in a low water system. By contrast, as $A_w$ increases above 0.7 the rate of NEB decreases due to dilution of reactants.

(2) **Solution model**. NEB takes place only in solution. The rate of NEB decreases with $A_w$ because the reactive solution volume fraction (the amount of solution phase in the product) decreases, leading to a net decrease in the rate of NEB.

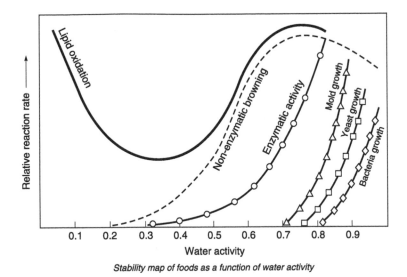

*Stability map of foods as a function of water activity*

**Figure 10.4**   Effect of water on deteriorative processes in food.

(3) **Glass transition model**. The emphasis is on molecular mobility and the effect of moisture as plasticizing agent for amorphous food products (Chapter 7). According to the glass transition scheme, the rate of food deterioration is controlled by the **microviscosity** of a food matrix. In a low moisture atmosphere, a food matrix exists in a glassy state with extremely high micro-viscosity. Within the glassy sate, most physicochemical changes and deterioration are slowed or negligible.

As moisture content is increased the microviscosity of a food matrix decreases in accordance with the **Gordon–Taylor equation** (Chapter 7, Section 7.4.5). The microviscosity also changes with temperature. The dependence of NEB rate on temperature is modeled by the **William–Ferry–Landel equation**, which summarizes changes in microviscosity of amorphous polymers with temperature:

$$\log \frac{\eta}{\eta_r} = \frac{A(T - T_r)}{B - (T - T_r)}, \tag{10.1}$$

where $T_r$ and $\eta_r$ are a reference temperature and microviscosity, respectively. The constant $A = -17.44$ and $B = 51.6$. At first, investigators assumed the $T_r$ was the same as the glass transition temperature, $T_g$. It was also assumed that both the diffusion rate for small reactive molecules *and* the generalized rate of food deterioration ($k$) would be inversely related to microviscosity. Therefore, a graph of $\log (k_r/k)$ versus $(T - T_g)$ should fit a straight-line graph. These predictions were confirmed experimentally.[29]

## 10.4.2   EXTRUSION COOKING

The effect of **extrusion cooking** on NEB was investigated using different temperature–time–moisture–pH combinations, by heating a mixture of wheat starch, glucose, and lysine in a 96:3:1 weight ratio. The mixtures were processed through a food extruder. The conditions of such experiments were chosen in order to mirror processing conditions used in industry including, extruder temperature of 136–145°C or 146–155°C (at the final barrel zone),

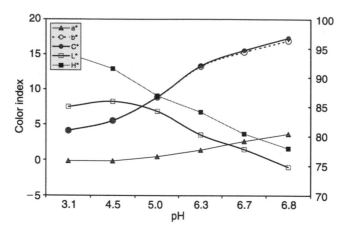

**Figure 10.5** Effect of pH on color formation during extrusion cooking (model starch system).

pH 3.1–6.8 (pH 3.1, 5.0 or 6.8), moisture (13, 15, 18%) and residence times of 32 and 45 s. The color of extruded wheat starch was determined using the CIE L* a* b* values. Over the range of variables tested, color was most affected by pH > temperature > moisture > time. The color of processed food products ranged from yellow (pH 3.1) to dark brown (pH 6.8). The results are consistent with the formation of different classes of MRPs under different conditions.[30,31] Clearly also the color indices b* and C* showed the greatest changes with processing pH. The effect of pH on color formation is shown in Figure 10.5.

### 10.4.3  Effect of pH on the Maillard Reaction

The Maillard reaction is pH sensitive. The extent of color formation increases with increasing pH. There are at least two possible reasons for the observed pH dependence: (1) alkaline conditions keep the amine group in an un-ionized and highly reactive state, and (2) alkaline conditions encourage the fragmentation of the Amadori intermediate (Figure 10.2).

## 10.5  CONCURRENT NONENZYMIC OXIDATION AND BROWNING

A survey of NEO and NEB reactions reveals important parallels. Both NEO and NEB involve (di)carbonyl compounds as key intermediates. For instance, lipid oxidation leads to **hydroperoxides**, which degrade to secondary products such as hexanal. The early part of the NEB also leads to (di)carbonyl compounds when **Amadori intermediates** decompose to deoxyosones. Carbonyl compounds, whether produced from NEO and NEB, then take part in **melanoidin** formation. Interrelations between NEO and NEB pathways are not wholly understood. The relative contribution of NEO and NEB to melanoidin formation is uncertain[b]. Some suggested links between lipid oxidation and enzymic browning are discussed below.

---

[b]Both NEO and NEB reactions have enzymic counterparts catalyzed by **oxidases**. Enzymic browning occurs when polyphenol oxidase catalyzes the oxidation of polyphenol to aromatic dicarbonyl compounds or quinones. These intermediates then undergo secondary NEB by reacting with amino acids, polypeptides or proteins. Enzymatic lipid oxidation is catalyzed by lipoxygenase, and leads to "fatty hyderperoxides" which then degrade to form volatile and nonvolatile carbonyl and dicarbonyl compounds. The former contribute to off-flavor. The latter can undergo NEB.

### 10.5.1 MELANOIDIN FORMATION BY NONENZYMIC OXIDATION AND BROWNING

A well-known feature of the Maillard reaction is the formation of the brown melanoidins. Recent studies have shown that products from lipid oxidation contribute to melanoidin formation.[32] Bovine serum albumin (BSA) was heated at 25–125°C for 24 h in the presence of a carbohydrate (ribose), linoleic acid hydroperoxide (13-hydroperoxy-9,11-octadecanoate) or hydroperoxide degradation products (decanal). In all cases fluorescent compounds and MRCs were formed. Indeed, the fluorescence spectra and other product characteristics were not really distinguishable. The reactions of BSA with ribose or hydroperoxide *degradation products* were incredibly similar with respect to (1) fluorescence emission spectra, (2) color change, (3) yellowness index, (4) temperature dependence, and (5) effect of pH on reaction extent. Results from these investigations[32] are partly summarized in Figures 10.6–10.9.

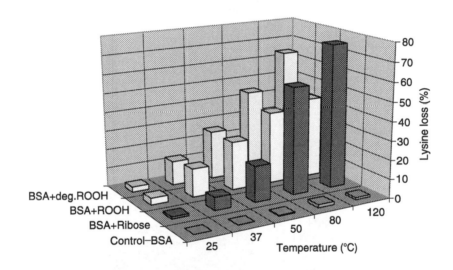

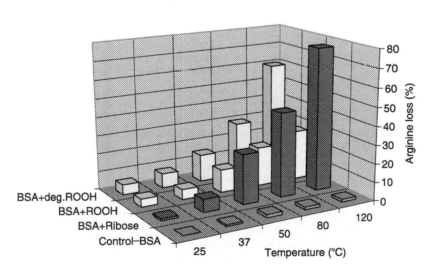

**Figure 10.6**  Effect of temperature on lysine (top) and arginine losses when BSA is incubated with ribose, lipid hydroperoxide (ROOH) or hydroperoxide degradation products (deg. ROOH) at pH 7. (Based on data calculated from Ref. 32–33.)

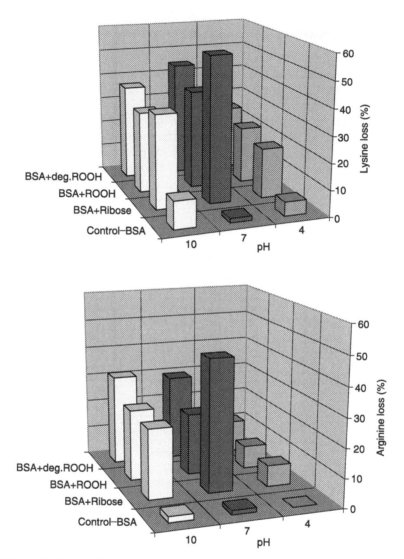

**Figure 10.7** Effect of pH on lysine (top) and arginine losses at 80°C when BSA is incubated with ribose, lipid hydroperoxide (ROOH) or hydroperoxide degradation products (deg. ROOH). (Based on data calculated from Ref. 32–33.)

Incubating BSA with lipid hydroperoxides leads to NEB. The products were virtually the same as those formed when BSA reacts with a simple sugar, ribose. There were increasing losses of free lysine and arginine with increasing *temperature* (Figure 10.6). The high losses of essential amino acids, up to 70–75% in some cases, are noteworthy. These figures probably underestimate of the real losses because some early MRPs (the Schiff's bases) degrade during amino acid analysis.

The *pH dependence* of color formation and the yellowness index were parallel to the losses of free lysine and the arginine (Figure 10.7). Browning, whether by carbohydrates or lipid oxidation products, was higher at pH 7 compared to pH 4. As described above, alkaline conditions (pH 10) encourage fragmentation of the Amadori product leading to color reactive intermediates. In contrast, neutral pH conditions favor the formation of 1-deoxyosone, which has both color and flavor forming capacity. A solvent of pH 4 or lower favors the formation of 3-deoxyosone and lower degrees of color (Figure 10.7).

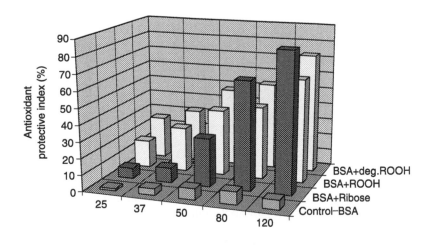

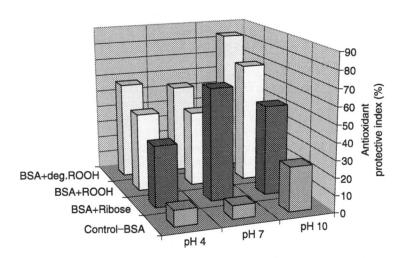

**Figure 10.8** Effect of temperature (top) and pH on the formation of antioxidant products when BSA is incubated with ribose, lipid hydroperoxide (ROOH) or hydroperoxide degradtion products (deg. ROOH). (Based on data calculated from Ref. 32–33.)

### 10.5.2 ANTIOXIDANT FORMATION BY NONENZYMIC OXIDATION AND BROWNING

A complementary relation applies to the formation of antioxidant products by NEO and NEB (Figure 10.8).[33] Thus, BSA was heated with ribose, linoleic acid 13-hydroperoxide or hydroperoxide degradation products, at 25–80°C (pH 7, pH 4 or pH 10) for up to 240 h. The pre-heated samples were cooled and then examined for antioxidant concentration using the TBA assay (Chapter 9). Measurements using BHT were used as a reference. The antioxidant capacity of protein samples was expressed as the BHT-equivalent **protective index** (PI):

$$PI(\%) = 100\left(\frac{TBARS_{sample} - TBARS_{BHT}}{TBARS_{Oil} - TBARS_{BHT}}\right), \tag{10.2}$$

where the PI(%) is calculated from the concentration of thiobarbituric acid reactive substances (TBARS) formed during a standard soybean oil oxidation test at 60°C, with or without the test sample or BHT. A PI value equal to 100% was an indication that the products were as efficient as BHT. This study confirmed that proteins react with sugars (BSA + ribose) to form antioxidant species. Protein reactions with lipid oxidation products also led to antioxidants.

# References

1. Hodge, J. E., Chemistry of browning reactions in model systems, *J Agricul. Food Chem.* 1, 928–943, 1953.
2. Ames, J. M., The Maillard browning reaction—an update, *Chem. Ind.* 5, 558–561, 1988.
3. Ames, J. M., Applications of the Maillard reaction in the food industry, *Food Chem.* 62(4), 431–439, 1998.
4. Labuza, T. P., Reineccius, G. A., Monnier, V. M., O'Brien, J., and Baynes, J. W., Eds, *Maillard Reactions in Chemistry, Food and Health*, Royal Society of Chemistry, Cambridge, UK, 1994.
5. O'Brien, J., Ames, J. M., Nursten, J. M., and Crabbe, J., *The Maillard Reaction in Foods and Medicine*, Royal Society of Chemistry, Cambridge, UK, 1998.
6. Rizzi, G. P., Chemical structure of colored Maillard reaction products, *Food Reviews Intern.* 13(1), 1–28, 1997.
7. Rizzi, G., The role of Maillard reaction in foods, in *Maillard Reactions in Chemistry, Food and Health*, Labuza, T. P., Reinesccius, G. A., Monnier, V., O'Brien, J. O., and Baynes, J. W., Eds, Royal Society of Chemistry, Cambridge, UK, 11–19, 1994.
8. Morales, F. J. and Boekel, M. A. J. S., A study on advanced Maillard reactions in heated casein/sugar solutions: color formation, *Intern. Dairy J.*, 8(10/11), 907–915, 1998.
9. Hoffman, T., Identification of novel colored compounds containing pyrrole and pyrrolinone structures formed by Maillard reactions of pentoses and primary amino acids, *J. Agricul. Food Chem.* 46(10), 3902–3911, 1998.
10. Hoffman, T., Studies on melanoidin-type colorants generated from the Maillard reaction of protein-bound lysine and furan-2-carboxaldehyde—chemical characterization of a red colored domain, *Zeitschrift fuer Lebensmittel Untersuchung und Forschung A/Food Research and Technology* 206(4), 251–258, 1998.
11. Ames, J. M., Dafaye, A. B., Bailey, R. G., and Bates, L., Analysis of the non-volatile Maillard reaction products formed in an extrusion-cooked model food system, *Food Chem.* 61(4), 521–524, 1998.
12. Lee, K. G. and Shibamoto, T., Toxicology and antioxidant activities of non-enzymatic browning reaction products: review, *Food Reviews Intern.* 18(2–3), 151–175, 2002.
13. American Institute of Cancer Research, Food nutrition and the prevention of cancer: a global perspective. Expert panel report, http://www.aicr.org/research/report_summary.lasso
14. Data from *The 1996 National Institute of Cancer Fact Book*, available on http://www3.cancer.gov/public/factbk96/fbindex.htm
15. Bartoszek, A., Mutagenic, carcinogenic, and chemopreventative compounds in foods, in *Chemical and Functional Properties of Food Components* (2nd edition), Sikorski, Z. E., Ed., CRC Press, New York, 307–336, 2002.
16. Goldman, R. and Shields, P. G., Food mutagens, *J. Nutri.* 133(3S), 965S–973S, 2003.
17. Schut, H. A. J. and Snyderwine, E. G., DNA adducts of heterocyclic amine food mutagens: implications of mutagenesis and carcinogenesis, *Carcinogenesis* 20(3), 353–368, 1999.
18. Ames, B. N., McCann, J., and Yamasaki, E., Methods for detecting carcinogens and mutagens with the *Salmonella*/mammalian-microsome mutagenicity test, *Mutat. Res.* 31, 347–364, 1975.
19. Maron, D. and Ames, B. N., Revised methods for the *Salmonella* mutagencity test, *Mutat. Res.* 113, 173–215, 1983.
20. Gee, P., Maron, D. M., and Ames, B. N., Detection and classification of mutagens: A set of base-specific *Salmonella* tester strains, *Proc. Natl. Acad. Sci. USA* 91, 11606–11610, 1994.

21. Rodriguez-Mellado, J. M. and Ruiz-Montoya, M., Correlations between chemical reactivity and mutagenic activity against *S. typhimurium* TA100 for alpha-dicarbonyl compounds as proof of the mutagenic mechanism, *Mutat. Res.* 304, 261–264, 1994.

22. Eisenbrand, G. and Tang, W., Food borne heterocyclic amines. Chemistry, formation, occurrence and biological activities. A literature review, *Toxicology* 84, 1–82, 1993.

23. Felton, J. S. and Knize, M. G., Heterocyclic-amine mutagens/carcinogens in foods, in *Handbook of Experimental Pharmacology*, Vol. 94/I, Cooper, C. S. and Grover, P. L., Eds, Springer-Verlag, Berlin, 471–502, 1990. Also see Felton, J. S. at the Lawrence Livermore National Laboratory website. http://www.llnl.gov/str/FoodSection3.html

24. National Cancer Institute, *Cancer Facts: Heterocyclic Amines in Foods*, 1996, available on http://cancer.gov/

25. Stavric, B., Biological significance of trace levels of mutagenic heterocycylic aromatic amines in human diet: a critical review, *Food Chem. Toxicol.* 32(10), 977–994, 1994.

26. Adamson, R. H. and Thorgeirsson, U. P., Carcinogens in foods: heterocyclic amines and cancer and heart diseases, *Adv. Exp. Med. Biol.* 369, 211–220, 1995.

27. Manzocco, L., Calligaris, S., Mastrocolar, D., Nicoli, M. C., and Lerici, C. F., Review of non-enzymatic browning and antioxidant capacity in processed foods, *Trends Food Sci.* 11, 340–346, 2000.

28. Labuza, T. P., Tannenbaum, S. R., and Karel, M., Water content and stability of low-moisture and intermediate-moisture foods, *Food Technol.* 24, 543–550, 1970.

29. Buera, M. P. and Karel, M., Application of the WLF equation to describe the combined effects of moisture and temperature on non enzymatic browning rates in food systems, *J. Food Process. Preserv.* 17, 31–45, 1993.

30. Sgaramella, S. and Ames, J. M., The development and control of color in extrusion cooked foods, *Food Chem.* 46, 129–132, 1993.

31. Nogochi, A., Mosso, K., Aymanrod, C., Jeunink, J., and Cheftel, J. C., Maillard reactions during extrusion cooking of protein-enriched biscuits, *Lebensm. Wiss. U. Technol.* 15, 105–110, 1982.

32. Hidalgo, F. J., Alaiz, M., and Zamora, R., Effect of pH and temperature on comparative nonenzymatic browning of proteins produced by oxidized lipids and carbohydrates, *J. Agricul. Food Chem.* 47, 742–747, 1999.

33. Alaiz, M., Hidalgo, F. J., and Zamora, R., Effect of pH and temperature on comparative antioxidant activity of nonenzymatically browned proteins produced by reaction with oxidized lipids and carbohydrates, *J. Agricul. Food Chem.* 47, 748–752, 1999.

# 11

# Food Enzymes

## 11.1  INTRODUCTION

Many changes that occur in fresh foods during storage are catalyzed by enzymes. Enzymatic activity affects many of the major food commodity groups: (1) fruits and vegetables, (2) cereals, legumes, and grains, (3) meat, poultry, and sea food, (4) beverages, including tea, chocolate, soft drinks, and wine, (5) milk and dairy products, (6) candy and confectionery. Preprocessed foods contain enzymes whose activity affects product quality in profound ways. Enzymes are also used in many food manufacturing processes. The production of cheese and high fructose syrup are examples of enzymatic food processes. Food enzymes can be divided into two classes: (1) **endogenous enzymes** are those that are naturally present in foods and which affect quality; and (2) **exogenous enzymes**, which are mostly isolated from bacteria, and are added to foods as **processing aids**. Papain isolated from papaya latex, and other plant proteases, are a notable exceptions to this rule.

This chapter introduces **food enzymes** and food enzymology.[1-3] With better understanding of enzymes, food technologists are able to develop newer and more efficient methods for food storage and processing. The remaining sections deal with enzyme kinetics (Section 11.2), the Michaelis–Menten equation (Section 11.3), enzyme inactivation for process endpoint determination (Section 11.4), isolation of enzymes (Section 11.5), and, finally, enzymes related to food quality (Section 11.6).

Enzymes are **biological catalysts**. Their role is to speed chemical reactions inside living cells (Eqs. 11.1a and 11.1b). Virtually all of the many hundreds of reactions taking place within living cells are enzyme catalyzed. Indeed, early scientists referred to living cells as bags of enzymes. One enzyme usually catalyses one reaction. The typical enzyme reaction is discrete and very simple (Eq. 11.1). Enzymatic reactions transform one chemical compound (substrate) into a second compound (product). The actual transformations will be familiar to most organic chemists; it seems that enzymes can be found in nature to perform most organic reactions, including esterification, hydrolysis, hydrogenation, isomerization, oxidation, oxygenation, reduction, polymerization, and many more.

$$\text{substrate (S)} \rightarrow \text{product(P)}, \tag{11.1a}$$

$$A + B \rightarrow C + D. \tag{11.1b}$$

Eq. 11.1b shows two substrates (A and B) and two products (C and D).

Enzymes are **highly specific** which means that they will recognize one type of substrate (or a group of closely related substrates) in the presence of hundreds of other compounds. Enzymes are highly **efficient** catalysts, able to work under very mild conditions of temperature, pH, and substrate concentration. To achieve phenomenal increases in reaction rate, enzymes lower the

**activation energy** for producing a transition. High **catalytic efficiency** requires strong enzyme binding to the substrate coupled with its rapid transformation on the enzyme surface.

According to current **theories of catalysis**, there are six possible reasons for the rate enhancement observed in the presence of an enzyme: (1) enzyme binding increases the strain within a substrate causing it to react, (2) enzyme binding with two substrates increases their effective concentration, (3) enzyme binding increases the collision frequency between two substrates, (4) enzyme binding improves the molecular orientation between reacting substrates and increases the yield of productive encounters between them, (5) general acid–base catalysis, and (6) enzymes have improved fit with the transition states and therefore encourage their formation.

## 11.2   ENZYME KINETICS

### 11.2.1   DEFINITIONS

Enzyme kinetics is the study of enzyme catalysed reaction *rates* and their dependence on factors such as enzyme concentration, substrate concentration, solvent pH, temperature, and ionic strength.

A principal goal in food enzymology is to measure the concentration of enzymes in food materials. Enzyme concentration is related to changes in the quality of stored foods. Since enzymes are invisible their concentration is determined by indirect means. Enzyme concentration is determined in terms of their "activity," normally expressed as the rate of product formation per unit time[a]. When determining enzyme concentration it is necessary to consider the effects of inhibitors and activators, as well as other environmental variables such as temperature and pH. In other words, enzyme concentration is directly proportional to enzyme activity. However, the constant linking activity with concentration changes with environmental conditions.

### 11.2.2   PRACTICAL CONSIDERATIONS FOR ENZYME ASSAYS

Measuring enzyme concentration, in terms of the observed activity, is a key task for food enzymologists. The concentration of enzymes in foodstuffs is related to the rate of deterioration. Furthermore, many processing endpoints (pasteurization, blanching, high pressure treatment, internal temperature of the product) can be assessed from the residual amount of enzyme present in a food. Finally, enzyme producers and users need a way to determine the amount of active enzyme present in any preparation.[4]

The list of some materials needed for enzyme experiments is shown in Table 11.1. The source of enzymes is usually a food material or the purified enzyme (Section 11.5). We rarely have the **natural substrate** for an enzyme. Therefore, a model reaction (Eqs. 11.1a and 11.1b) employing artificial substrates is used to monitor the rate of reaction. The goal is to measure the reaction *rate* in terms of the amount of product, $P$ (in moles) formed per unit time.

A variety of **synthetic substrates** are available from commercial suppliers for assaying all the six major enzyme groups: oxidoreductases, transferases, hydrolases, lyases, isomerases and ligases. The names reflect the type of reaction catalyzed. Artificial substrates are usually colorogenic chemicals, which break down in the presence of an enzyme to form colored products. The concentration of product is easily measured using absorbance or fluorescence

---

[a]Enzymatic activity is different from "activity" as used in connection with general chemistry. For example the activity of sodium chloride is determined by multiplying the concentration ($C$) by the activity coefficient ($\gamma$): activity $= \gamma C$. We also use the term activity in connection with moisture relations in foods. Water activity is the ratio of the water vapor pressure over a solution divided by the water vapor pressure over a pure solvent, at a specified temperature.

**Table 11.1**
**Material for conducting enzyme assays**

| Material | Comment |
|---|---|
| Source of enzyme | Food homogenate or extract, microbial fermentation broth |
| Substrate | Natural materials (e.g., starch, lipid or protein—casein), large variety of synthetic substrates available |
| Product detection method | Most popular being colorimeter or fluorescence, but extends to all analytical methods including HPLC, GLC, oxygen electrodes, and sensory analysis |
| Controlled solution | Mostly buffered solutions are used for pH control. pH-stat is used with fermenters |
| Temperature controller | Water bath, water jacketed vessels |
| Reaction vessels | Ranging from glassware and test-tubes to large batch or flow through reactors |
| Pipettes, glass cuvettes | |

measurements. Though convenient and relatively inexpensive, synthetic substrates may give the wrong impression about enzyme activity. Some natural substrates for enzymes such as proteases, lipases, and carbohydrases (protein, lipids, and starch) are also available for use.

The solvent used for an enzyme reaction can be precisely controlled with respect to composition, temperature, pH, and ionic strength. This is easily achieved through the use of standard **buffer solutions** to dissolve enzyme and substrate. The choice of buffer should be such that the required $pK_a$ ($-\log K_a$) for the buffer salts is no more that $\pm 1$ pH unit from the required pH. Buffer interactions with metal ions are another consideration. Phosphate will bind $Ca^{2+}$ ions and should be avoided if the enzyme being studied requires $Ca^{2+}$ for activity. Sodium azide, a well known preservative, can be added to buffers to stop microbial growth. However, azide is also a metal chelator, which inactivates copper containing enzymes such as polyphenol oxidase (PPO). The **temperature coefficient** of different buffers ($\theta$) shows the pH change obtained for a unit change in temperature. Phosphate buffer ($\theta = -0.0028\,pH/°C$) is preferable to Tris buffer ($\theta = 0.031\,pH/°C$) when it is necessary to assay an enzyme at a range of temperature conditions.

Despite the need for controlled conditions, we usually do not know the best solvent composition needed for a particular enzyme. Optimum assay conditions (buffer pH, temperature, requirement for specific metal ions, etc.) need to be established by trial and error. One strategy is to start with a set of solution conditions reported in the literature for an enzyme belonging to the same family as the one of interest. This will provide a rough guideline only. As an illustration, PPO from potatoes is relatively well studied. The enzyme can be readily assayed using 4-methyl catechol as substrate in a buffer with pH 7. By comparison the PPO isolated from taro showed the highest activity at pH 4.6.

### 11.2.3 THE MEASUREMENT OF INITIAL RATE

By agreed convention, enzymologists measure **initial rate** ($v_0$) for enzyme reactions. This is the rate of reaction at a time ($t$) when 5% or less of the substrate is converted to product (Eqs. 11.1a and 11.1b). To determine $v_0$ ($= dP/dt$), the concentration of product (P) should not exceed 5% of the starting substrate (S) concentration ($[P] < 0.05\,[S]$). Using $v_0$ measurements for enzyme studies avoids potential complications arising from excessive product build-up. Some enzymes are inhibited by their product. Also, without the initial rate restriction, the substrate concentration [S] becomes difficult to define because it changes continually during an enzyme reaction. Apart from Section 11.3.2.2 all discussions presented in this chapter are based on initial rate values.

To determine $v_0$ consider an enzyme reaction leading to a colored product. The rate of product formation ($dP/dt$) is conveniently measured using calorimetry in terms of the absorbance ($A$) increase over a timed period, $t$ (in seconds).

$$v_0 = \frac{dP}{dt} = \frac{Ar}{t\varepsilon l}.$$ (11.2)

In Eq. 11.2, the reaction volume is $r$ (in liters), $\varepsilon$ (liter/mole/cm) is the molar extinction coefficient and $l =$ the cuvette size (cm). The preceding calculation gives $v_0$ as moles of product formed per minute. Doubling or halving the reaction time should give the same numerical value $v_0$, otherwise the rate measurement does not qualify as the initial rate.

## 11.3 MICHAELIS–MENTEN EQUATION

Enzymes combine with the substrate at the **active site**. This a discrete region on the enzyme protein loaded with the amino acid groups needed to perform catalysis. The enzyme–substrate complex then goes through a reaction cycle, which transforms the substrate(s) into a finished product (P). The **enzyme–product** complex now dissociates freeing the enzyme for another cycle of catalysis (Figure 11.2). As more substrate molecules are added, the rate of product formation ($v_0$) rises until each enzyme molecule is fully **saturated**. The idea of enzyme saturation is accredited to Adrian Brown.[5]

As the substrate concentration is increased, for a fixed concentration of enzyme, $v_0$ gradually approaches the maximum reaction rate possible, $V_{max}$:

(1) The maximum rate is determined by the number of enzyme molecules present, [$E_T$].
(2) $V_{max}$ is dependent on the time for one enzyme molecule to transit through one catalytic cycle. The inverse transit period is called the enzyme **turnover number** ($k_{CAT}$).
(3) The magnitude of $k_{CAT}$ shows the number of substrate molecules converted to product, per second. Some of the highest values for $k_{CAT}$ have been recorded for urease, catalase, and carbonic anhydrase.

Relations between $v_0$ and [S] are described by the Michaelis–Menten equation (11.3). Figure 11.1 shows Professor Maud Lenore Menten, a key person in the development of enzyme kinetics. Figure 11.2 shows the enzyme–substrate reaction scheme.

$$v_0 = \frac{V_{max}S}{K_m + S},$$ (11.3)

where $S =$ substrate concentration (mole/li), $V_{max} =$ theoretical maximum velocity for an enzyme reaction (μmole/s) and $K_m$ (mole/li) is called the Michaelis–Menten constant.

### 11.3.1 MICHAELIS–MENTEN PARAMETERS, $V_{MAX}$ AND $K_M$

#### 11.3.1.1 Significance of $V_{max}$

$V_{max}$ is the maximum reaction of rate possible when a *fixed* amount of enzyme is exposed to an infinitely high concentration of substrate. $V_{max}$ is determined by the total enzyme concentration ($E_T$):

$$V_{max} = k_{CAT}[E_T].$$ (11.4)

**Figure 11.1** Professor Maud Lenore Menten (1879–1960) a key person in the development of enzyme kinetics. Maud Lenora Menten was one of the earliest Canadian women to receive a doctorate. She obtained her BA in 1904 and a master's degree in biology from the University of Toronto in 1907. Her medical doctorate was also from the University of Toronto and followed in 1911. The Michaelis–Menten equation was formulated during Dr Menten's studies in Berlin (1913) with Lenoir Michaelis. Maud Menten also earned a PhD in biochemistry at the University of Chicago (1916) and later became an assistant professor (1923) on the faculty of the University of Pittsburgh School of Medicine where she worked until her retirement in 1950. (Information adapted from a memorial at the University of Toronto). (Photograph from: A picture gallery of female scientists, www:imp.leidenuniv.nl/~schmidt/wome-nimg.html)

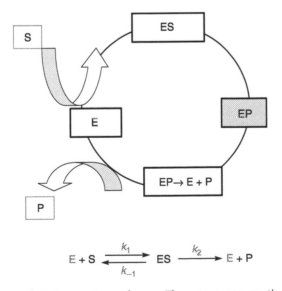

**Figure 11.2** The enzyme–substrate reaction scheme. The rate constants $(k_1, k_2, k_3)$ are for the steps shown. The constant $k_2$ is more commonly called the turnover number $(k_{CAT})$. The idea of enzyme saturation is accredited to Adrian Brown.

As described previously, $k_{CAT}$ is a constant of proportionality also called the turnover number. For practical applications the concentration of enzyme is measured in terms of a $V_{max}$ scale described as the **Katal**, which is the quantity of enzyme sufficient to catalyze the formation of 1 mole of product per second (Eqs. 11.5a–c)[b].

$$Katal = 1 \text{ mole (of product formed) per second,} \tag{11.5a}$$

$$IU = 1 \, \mu mole \text{ (product) per minute,} \tag{11.5b}$$

$$IU = 60 \, \mu Katal. \tag{11.5c}$$

The other popular measure of enzyme concentration is the **international unit** (IU). One IU of enzyme activity is the amount of enzyme sufficient to produce $1 \times 10^{-6}$ mole of product per minute. Enzyme prices are normally based on their **specific activity**, which is the number of IU of activity displayed per unit weight of enzyme preparation. For a given source material, the specific activity is $V_{max}$ divided by the general concentration of protein:

$$\text{specific activity} = \frac{\text{enzyme activity}}{\text{sample weight}} = \frac{V_{max}}{\text{mg protein}}. \tag{11.6}$$

### 11.3.1.2 The Michaelis–Menten Constant

$K_m$ is defined as the substrate concentration necessary for $v_0$ to reach 50% $V_{max}$. Therefore, $K_m$ is an indirect measure of **enzyme affinity** for the substrate. For any two enzymes, the one possessing a low $K_m$ value has greater affinity for its substrate. A large $K_m$ value indicates low substrate binding affinity. The value for $K_m$ is constant for any enzyme. Tabulated values for $K_m$ are available to help differentiate between related enzymes. The relation between the $K_m$, [S], $V_{max}$ and $v_0$ is summarized by the Michaelis–Menten equation (Eq. 11.3). With a high concentration of substrate, relative to $K_m$ (mole/l), $v_0$ approaches $V_{max}$. Under conditions when $[S] = 2K_m$, $[S] = 5K_m$ or $[S] = 10K_m$, we can show from Eq. 11.3 that $v_0 = 0.66 \, V_{max}$, $v_0 = 0.83 \, V_{max}$ and $v_0 = 0.91 \, V_{max}$, respectively. A hypothetical enzyme having a $K_m$ value of 1–10 mM requires a substrate concentration of 100–1000 mM, which is 10–100 $K_m$ "equivalents" of substrate, in order for $v_0$ to approach 99% of $V_{max}$.

The effects of **low** concentrations of the substrate on the rate of reaction can be also anticipated from the Michaelis–Menten equation (Table 11.2). For instance, at low concentrations of substrate compared to $K_m$ ($[S] < K_m$) then $v_0 = V_{max} [S]/K_m$. In contrast, at substrate concentration "equivalent" to $K_m$, Eq. 11.3 becomes $v_0 = \frac{1}{2}V_{max}$. This last result means that $K_m$ is the substrate concentration needed for an enzyme reaction rate to proceed at 50% $V_{max}$. Interestingly, $v_0$ is *always* dependent on $V_{max}$. In other words, enzyme activity is always dependent on enzyme concentration almost regardless of the substrate availability (Table 11.2). For this reason enzyme studies always employ a fixed concentration of enzyme. Of course, enzymes cannot work if the substrate concentration is zero.

---

[b]The unit Katal (pronounced "cattle") was introduced by the Système International d'Unités (SI) in October 1999. See http://www.unc.edu/~rowlett/units/sipm.html for further information.

---

**Table 11.2**
**Substrate and $K_m$ effects on Michaelis–Menten kinetics**

| Substrate concentration | Michaelis–Menten equation |
|---|---|
| High, $[S] > K_m$ | $v_0 = \dfrac{V_{max}[S]}{[S]} = V_{max}$ |
| Low, $[S] < K_m$ | $v_0 = \dfrac{V_{max}[S]}{K_m} = k_1[S]$ |
| Equivalent, $[S] = K_m$ | $v_0 = \dfrac{V_{max}[S]}{2[S]} = \dfrac{V_{max}}{2}$ |

---

**Table 11.3**
**Evaluating Michaelis–Menten parameters by graphical analysis**

| Equation of the graph | x-axis | y-axis | Gradient | Intercept |
|---|---|---|---|---|
| **Lineweaver–Burk** $\dfrac{1}{v_0} = \dfrac{K_m}{V_{max}[S]} + \dfrac{1}{V_{max}}$ | $\dfrac{1}{[S]}$ | $\dfrac{1}{v_0}$ | $K_m/V_{max}$ | $1/V_{max}$ |
| **Eadie plot** $v_0 = V_{max} - \dfrac{v_0 K_m}{[S]}$ | $\dfrac{v_0}{[S]}$ | $v_0$ | $-K_m$ | $V_{max}$ |
| **Hanes plot** $\dfrac{[S]}{v_0} = \dfrac{K_m}{V_{max}} + \dfrac{[S]}{V_{max}}$ | $[S]$ | $\dfrac{[S]}{v_0}$ | $1/V_{max}$ | $K_m/V_{max}$ |

---

## 11.3.2 Evaluating Michaelis–Menten Parameters

The two Michaelis–Menten parameters ($V_{max}$ and $K_m$) provide a great deal of useful information about different enzymes. For instance, enzyme producers usually determine $V_{max}$ for quality control during enzyme production. The enzyme purchaser uses $V_{max}$ to assess the potency of different batches of enzyme reagents and activity loss over time. Knowledge of $K_m$ allows us to design enzymatic reactions that proceed at some fraction of $V_{max}$. Analysis of inhibition and activator effects requires knowledge of $K_m$ and $V_{max}$ with and without inhibitor or activator present. The most common ways to determine $V_{max}$ and $K_m$ are explained below.

### 11.3.2.1 Double Reciprocal and Other Graphical Methods

It is not convenient to measure $V_{max}$ directly because $v_0$ approaches $V_{max}$ only when $[S] = 100\,K_m$. However, most enzyme substrates are either too expensive and/or too difficult to dissolve at the high concentrations needed to produce a $100\,K_m$ solution. Furthermore, exposing enzymes to high concentrations of substrate can lead to **substrate inhibition** arising from non-productive encounters between the enzyme and substrate. In practice, $V_{max}$ and $K_m$ are usually determined from a limited number of $v_0$ and $[S]$ data, using linear forms of the Michaelis–Menten equation (Table 11.3).

To determine the values of $K_m$ and $V_{max}$ for a specific enzyme, we require $v_0$ measurements $[v_{0,1}, v_{0,2}, \ldots, v_{0,6}]$ recorded over a *range* of different substrate concentrations, $[S_1, S_2, \ldots, S_6]$.

For all measurements the concentration of enzyme and all other experimental conditions (Table 11.1) are kept constant. Only the substrate concentration is changed for different $v_0$ measurements. The final data is a table of values for $v_0$ and the corresponding [S] values. A method of data analysis is now required that enables us to extract $V_{max}$ and $K_m$ values from the set of $v_0$ measurements recorded at different substrate concentrations.

The first graphical method for estimating $K_m$ and $V_{max}$ employs the **Lineweaver–Burk** *equation* (Table 11.3). This is a straight line equation with the form, $y = mx + C$. Plotting $1/[S]$ (*x*-axis) against $1/v_0$ (*y*-axis) will produce a straight-line graph with a gradient $= K_m/V_{max}$ and the $x = 0$ intercept $= 1/V_{max}$. There are two other linear graphs for estimating $V_{max}$ and $K_m$ values from intitial rate measurements (Table 11.3). The **Eadie equation** is obtained by multiplying the Lineweaver–Burk equation by both $v_0$ and $V_{max}$. Similarly multiplying the Lineweaver–Burk plot by [S] gives the **Hanes plot**. Linearization strategies for the Michaelis–Menten equations are shown in Figure 11.3.

The three common graphical methods for estimating $K_m$ and $V_{max}$ have their advantages and disadvantages. The Lineweaver–Burk plot is the most popular method. One reason for this is that it is most likely to yield a straight-line graph. The disadvantage of the Lineweaver–Burk plot is that it uses inverse values for the data ($1/[S]$ and $1/v_0$). Taking inverse values gives undue emphasis to rate measurements recorded at low substrate concentrations where experimental error is liable to be greatest. A complete analysis of the merits of each the linear equation in Table 11.3 is beyond the scope of this book and the interested reader is referred to enzyme kinetics monographs by Marmase[6] and Engels.[7]

### 11.3.2.2 Integrated Michaelis–Menten Equations

Industrial enzyme processes achieve high levels of substrate conversion. For economic reasons, it is usually desirable to convert greater than 5% of substrate to product. **Non-initial rate**

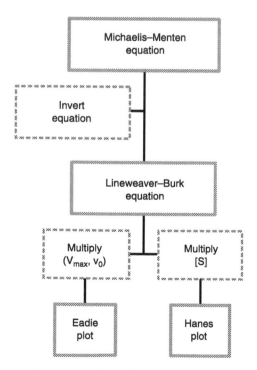

**Figure 11.3** Linearization strategies for the Michaelis–Menten equation (see text for details).

conditions are encountered during most industrial applications. Initial rate conditions are neither enforceable nor desirable. Initial rate kinetics do not apply to most practical enzyme reactions. The **integrated Michaelis–Menten** equation (Eqs. 11.7 and 11.8) provides a means of analyzing enzyme reactions under non-initial rate conditions.

$$\frac{P}{t} = V_{max'} - \frac{K_{m'}}{t} \ln\left(\frac{[S]_t}{[S]_0}\right), \tag{11.7}$$

$$P = V_{max'}t - K_{m'} \ln(1 - X), \tag{11.8}$$

where $[S]_0$ and $[S]_t$ are substrate concentrations for times $t = 0$ and $t =$ any time; the concentration of product is $P$ $(= [S]_0 - [S]_t)$ and $X (= P/[S]_0)$ is the fraction of substrate converted to product.

Values for the apparent Michaelis–Menten constant ($K_{m'}$) and maximum velocity ($V_{max'}$) can be extracted by monitoring the product concentration as function of time. A graph of $[P]/t$ vs $(1/t) \ln([S]_t/[S]_0)$ should be a straight line with a gradient equal to $- K_{m'}$ and an intercept of $V_{max'}$.

### 11.3.3 DERIVATION OF THE MICHAELIS–MENTEN EQUATION

Enzymologists have two ways of deriving the Michaelis–Menten equation, which is to say, two ways for accounting for Eq. 11.3. Understanding how the Michaelis–Menten equation was originally formulated will show some of its underlying **assumptions**.

The Michaelis–Menten equation is based on the enzyme–substrate cycle shown in Figure 11.2 proposed by Fischer in 1902. From Figure 11.2, it may be assumed that:

(1) An enzyme mass balance can be drawn up. According to Eq. 11.9 enzyme is present in the free state (E) or as an enzyme–substrate complex (ES).
(2) The concentration of enzyme is much smaller than substrate.
(3) The initial rate depends on the *concentration* of ES formed and the rate for its breakdown, $k_{CAT}$ (see Eq. 11.10).
(4) In the presence of a **saturating** concentration of substrate all enzyme molecules exist as ES.

It follows from this last assertion and Eq. 11.4 that $V_{max}$ can be described by Eq. 11.11.

$$E_T = E + ES, \tag{11.9}$$

$$v_0 = k_{CAT}[ES], \tag{11.10}$$

$$V_{max} = k_{CAT}[ES]. \tag{11.11}$$

The preceding relations, and the underlying assumptions, are key to understanding the Michaelis–Menten equation. For instance, the *specific* rate reaction ($v_0/[E_T]$) can be simply expressed by dividing Eq. 11.10 by Eq. 11.9. The resulting relation (Eq. 11.12) shows remarkable resemblance to the Michaelis–Menten equation (Eq. 11.3):

$$\frac{V_0}{[E_T]} = \frac{k_{CAT}[ES]}{E + [ES]}. \tag{11.12}$$

We proceed from Eq. 11.12 to the full Michaelis–Menten equation using a number of other assumptions about the behavior of enzymes as detailed by the rapid equilibrium model (Section 11.3.3.1) or steady-state model (Section 11.3.3.2).

### 11.3.3.1 The Rapid Equilibrium Model

The rapid-equilibrium model of Michaelis and Menten assumes that enzymes bind extremely rapidly with their substrate, leading to equilibrium ($E + S \rightleftharpoons ES$). It is after this equilibrium is established that ES breaks down *slowly* to the free enzyme and product. Therefore an *equilibrium* dissociation constant ($K_s$) can be defined for an enzyme reaction (Eq. 11.13).

$$K_s = [E][S]/[ES] \tag{11.13}$$

In Eq. 11.12, the term [ES] is not easily measurable. Therefore, Michaelis and Menten subtituted $[ES] = [E][S]/K_s$ into Eq. 11.12 and produced Eq. 11.14, from which it is easy to show that Eq. 11.3 applies[c].

$$\frac{v_0}{[E_T]} = \frac{k_{CAT}[E][S]/K_s}{[E] + [E][S]/K_s}, \tag{11.14a}$$

$$v_0 = \frac{V_{max}[S]}{[K_s] + [S]}. \tag{11.14b}$$

### 11.3.3.2 The Briggs–Haldane Steady-State Model

Briggs and Haldane noted that the rapid equilibrium assumption was too restrictive. They proposed that the enzyme–substrate catalytic cycle (Figure 11.2) is likely to produce a steady-state concentration for ES. That is, the rate of formation of ES and its degradation would be equal.

$$\text{Rate of ES formation} = \text{rate of ES degradation.} \tag{11.15a}$$

$$k_1[E][S] = (k_{CAT} + k_{-1})[ES]. \tag{11.15b}$$

As a consequence of these assertions, they showed that $[ES] = k_1[E][S]/(k_{CAT} + k_{-1})$ or $[ES] = [E][S]/K_m$. Briggs and Haldane then substituted for [ES] in Eq. 11.12 as before. Starting from Eq. 11.12 and substituting for [ES] ($= [E][S]/K_m$) they produced Eq. 11.3 as before. Comparing the rapid equilibrium and steady-state derivation we can see that

$$K_m = \frac{k_{CAT} + k_{-1}}{k_1}, \qquad K_s = \frac{k_{-1}}{k_1}. \tag{11.16}$$

Michaelis and Menten assumed that $K_m$ (more correctly $K_s$) was an equilibrium (dissociation) constant related to enzyme affinity for the substrate. In contrast, Briggs and Haldane showed that $K_m$ is not a dissociation constant *unless* $k_{-1} \gg k_{CAT}$. Without independent proof that $k_{-1} \gg k_{CAT}$, the "true" $K_m$ value defined by Briggs and Haldane is not a dissociation constant and $K_m$ is not an index of enzyme–substrate binding affinity.

---

[c]To produce the Michaelis–Menten equation from Eq. 11.14, examine the right hand side of this expression. First delete [E] from the numerator and denominator. Next, multiply top and bottom with $K_s$. Finally, replace ($k_{CAT}$ [$E_T$]) with $V_{max}$.

### 11.3.4 Enzyme Inhibition and the Inhibition Equations

An *enzyme* **inhibitor** is any small compound that, by interacting with a localized region on an enzyme molecule, depresses its activity. **Denaturants** such as urea and guanidine hydrochloride also reduce enzyme activity. Denaturants are not considered true inhibitors because they interact with the whole enzyme molecule rather than a specfic binding site. For the same reason, loss of enzyme activity due to **heat treatment** is not considered a form of inhibition. Inhibition patterns can provide useful information about enzyme structure. Inhibitors also reduce the efficiency of enzyme reactions. Techniques for assessing the types and strength of enzyme inhibition are considered in this section.

#### 11.3.4.1  Enzyme Reversible Inhibition

The **reversible inhibitors** bind to enzymes by noncovalent bonding. This type of inhibition can be removed by dialyzing the enzyme sample to remove the inhibitor. By contrast, **irreversible** *inhibitors* bind to enzymes via covalent bonding. The effects of an irreversible inhibitor cannot be reversed by sample dialysis. An irreversibly inhibited enzyme can be reactivated by chemical means. Only reversible inhibition is amenable to Michaelis–Menten analysis. Three classes of reversible inhibition can be distinguished from their effects on values $K_m$ and $V_{max}$. These are **competitive, uncompetitive** or **noncompetitive** inhibition (Table 11.4).

A **competitive inhibitor** (CI) has a structure similar to the true substrate. Due to the close resemblance between a CI and the substrate the two molecules bind to the same enzyme active site. Therefore, a CI reduces enzyme affinity for the substrate. The $K_m$ for an enzyme substrate (strictly we should refer to the $K_s$) increases in the presence of a CI. Another possible mode of action for CI is to bind a different site on the enzyme causing a structural distortion that prevents the normal mode of enzyme–substrate binding. No matter the exact mechanism operating, the CI and substrate cannot bind to an enzyme at the same time.

#### 11.3.4.2  Inhibition Equations

The effect of CI on Michaelis–Menten kinetics can be explained using Figure 11.2. In addition to the ES complex, we must now consider enzyme binding with the CI. Enzyme binding with

**Table 11.4**
**Forms of reversible inhibition and effects on kinetic parameters**

| Type of inhibition | $K_m$ | $V_{max}$ | Complex formed | Lineweaver–Burk equation |
|---|---|---|---|---|
| None | – | – | – | $\dfrac{1}{v_0} = \dfrac{K_m}{V_{max}[S]} + \dfrac{1}{V_{max}}$ |
| Competitive | ↑ | – | EI | $\dfrac{1}{v_0} = \dfrac{K_m(1 + [I]/K_i))}{V_{max}[S]} + \dfrac{1}{V_{max}}$ |
| Uncompetitive | ↓ | ↓ | ESI | $\dfrac{1}{v_0} = \dfrac{K_m}{V_{max}[S]} + \dfrac{(1 + ([I]/K_i))}{V_{max}}$ |
| Noncompetitive | – | ↓ | EI and ESI | $\dfrac{1}{v_0} = \dfrac{K_m(1 + ([I]/K_i))}{V_{max}[S]} + \dfrac{(1 + ([I]/K_i))}{V_{max}}$ |

CI involves an equilibrium reaction (Eq. 11.17a) for which a dissociation constant ($K_i$) can be defined according to Eq. 11.17b.

$$E + I \rightleftharpoons EI, \qquad (11.17a)$$

$$K_i = [EI]/[E][I]. \qquad (11.17b)$$

In the presence of a CI, the Michaelis–Menten equation is remodeled using the rapid equilibrium derivation from Section 11.3.3.1. The final equation for CI is as shown in Eq. 11.18.

$$v_o = \frac{V_{max}[S]}{K_s(1 + ([I]/K_i)) + [S]}. \qquad (11.18)$$

This relation is the same as Eq. 11.3 but $K_s$ is multiplied by $(1 + ([I]/K_i))$. (See also Appendix 11.1.) Eq. 11.19 shows the corresponding Lineweaver–Burk relation[d]. Eq. 11.19 shows how an enzyme *ought* to behave when exposed to a competitive inhibitor. This response and other relations describing **uncompetitive** and **non-competitive** inhibition are compared in Table 11.4.

$$\frac{1}{v_0} = \frac{K_m(1 + ([I]/K_i))}{V_{max}[S]} + \frac{1}{V_{max}}. \qquad (11.19)$$

Inhibition equations are useful for identifying different mechanisms of enzyme inhibition (Table 11.4). Figure 11.4 shows the effect of a competitive inhibitor on the Lineweaver–Burk plot. With a CI present, the *observed* Michaelis–Menten constant ($K_{m*}$) is increased by the factor $(1 + ([I]/K_i))$ compared to the $K_m$ value *in the absence of inhibitor*. $K_{m*}$ increases with increasing concentrations of the inhibitor. The measured $V_{max}$ value in the presence of a CI is unchanged compared to the "true" value.

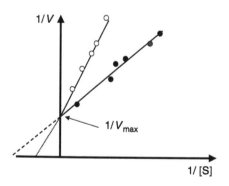

**Figure 11.4** Lineweaver–Burk plot for an enzyme without (closed circles) and with a competitive inhibitor (open circles).

---

[d]The preceding equations contain the variable $K_s$ as the Michaelis–Menten constant. Whilst this is strictly the correct terminology most books dealing with reversible inhibition express the Michaelis–Menten constant as $K_m$. For the remainder of this discussion, we will now "switch" to using $K_m$, though this is not strictly correct in the context of reversible inhibition.

Two scenarios requiring the use of enzyme inhibitor equations will be considered. First, a crude food enzyme extract may contain reversible inhibitors. Under such circumstances, the measured $K_m$ and $V_{max}$ values may not be true values. To avoid errors arising from reversible inhibition the crude enzyme samples can be "cleaned up" by dialysis. In general values for $K_m$ or $V_{max}$ may change during enzyme purification.

The inhibitor equations are also useful for evaluating the **strength** *of inhibition* measured by the magnitude of $K_i$. Briefly, measurements of the initial rate are recorded for a fixed concentration of enzyme *and* added inhibitor, $[I]_1$. $v_0$ is measured for a range of substrate concentrations. A control study is performed in the same manner except that no inhibitor is added. A Lineweaver–Burk graph is constructed of $1/v_0$ plotted versus $1/[S]$. From the pattern of change for $K_m$ and $V_{max}$ values we can identify the reversible inhibitor as either competitive, uncompetitive or noncompetitive. A graph such as shown in Figure 11.3, identifies the added inhibitor as a CI. For more reliable results, repeat the above analysis at several concentrations of added inhibitor, $[I]_2$, $[I]_3$ etc. The value of the inhibitor constant, $K_i$ is then determined as follows. For a competitive inhibitor the value of $K_{m*}$ changes with inhibitor concentration according to Eqs. 11.20a and 11.20b.

$$K_{m*} = K_m(1 + ([I]/K_i)) \tag{11.20a}$$

$$K_{m*} = K_m + K_m([I]/K_i) \tag{11.20b}$$

Drawing a graph of $[I]$ versus $K_{m*}$ will yield a straight line with a gradient equal to $K_m/K_i$ and an $x = 0$ intercept equal to $K_m$. Enzymes subject to uncompetitive or noncompetitive inhibition can also be analyzed from the principles outlined above.

## 11.3.5   Effect of pH on Enzyme Activity

Enzyme reactions are sensitive to pH due to one of the following reasons: (1) the enzyme reaction involves a hydrogen ion ($H^+$) as substrate or product; (2) one or more substrates undergoes ionization thereby altering the enzyme–substrate binding affinity and/or $V_{max}$; (3) the enzyme protein denatures at extremes of pH; and (4) one or more enzyme active site groups ionizes at a different pH.

Michaelis–Menten kinetics is readily applied to the active site effect. According to the most popular description of pH effects, changes occur in enzyme activity due to an active site group undergoing ionization. Each enzyme active site is assumed to possess one such ionizable group. The active site group binds with $H^+$ ions as defined by the association constant ($K_a$).

$$E + H^+ \rightleftharpoons EH. \tag{11.21a}$$

$$K_a = [EH]/[E][H^+]. \tag{11.21b}$$

Enzyme binding to $H^+$ results in enzyme inhibition or activation. When a $H^+$ ion functions as a **reversible inhibitor** the active enzyme species is E. Alternatively, where $H^+$ functions as a **reversible activator** then EH is the active form of enzyme.

Active site groups can be identified from the effect of pH on $V_{max}$. Generally, a graph of $V_{max}$ versus pH is S-shaped with an inflexion point where $V_{max}$ changes most rapidly as a function of the pH. The inflexion point on a $V_{max}$–pH graph occurs when the solvent pH equals the p$K_a$ ($= -\log K_a$) value for the enzyme active site group. Alternatively, drawing a graph of $\log V_{max}$ versus pH shows two simple straight-line segments that intersect at the p$K_a$. Amino acid side chains have characteristic p$K_a$ values. For example, an enzyme

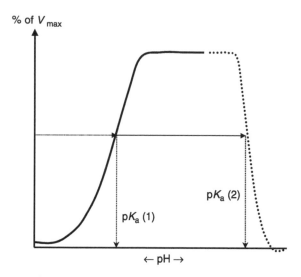

**Figure 11.5**   Activity pH profile for an enzyme having two ionizable groups at the active site. The sites are treated independently of each other. The plateau shows the optimum pH for activity.

possessing an active site aspartate residue has a $pK_a$ of 3.0–4.0. An active serine residue leads to $pK_a \sim 6.8$. Sulfhydryl enzymes with cysteine at their active site have $pK_a = 8$–8.5. Some uncertainty arises because the $pK_a$ values of amino acid side chains change depending on their surroundings. The activity pH profile for an enzyme with two ionizable groups is shown in Figure 11.5.

### 11.3.6  Effect of Temperature on Enzyme Activity

The effect of changing temperatures on enzymes is two-fold. With every 10°C rise in temperature $v_0$ rises two-fold. However, denaturation usually sets in above a certain temperature. The combination of rising activity and denaturation produces a bell-shaped temperature–activity profile. The temperature dependence of $V_{max}$ can be expressed by the **Arrhenius equation:**

$$\ln V_{max} = \ln A - \Delta E^{\#}/RT. \tag{11.22}$$

Therefore, a graph of $\ln V_{max}$ versus $1/T$ (where $T$ is in kelvin) yields a straight-line graph with a slope of $-\Delta E^{\#}$ which is the activation energy for an enzyme catalyzed reaction. In Eq. 11.22, the graph intercept ($\ln A$) is related to molecular orientation and collision frequency of reacting species. Recall that $V_{max}$ ($=k_{CAT}$ [$E_T$]) is a rate parameter and that the Arrhenius equation describes how reaction rates vary with temperature (Chapter 1, Eq. 1.2).

    The temperature dependence of $K_m$ (or $K_s$) yields information about enzyme–substrate binding. Recall that $K_s$ is a dissociation constant for the binding interaction leading to the enzyme–substrate complex (E + S $\rightleftharpoons$ ES). From physical chemistry principles, $1/K_s$ changes with temperature as described by the van't Hoff equation, $\ln (1/K_s) = C - \Delta H_s/RT$. Therefore, a graph of $\ln (1/K_s)$ versus $1/T$ (K) will produce a straight line with a gradient equal of $-\Delta H_s$ which is the enthalpy change for enzyme–substrate binding. The Gibbs free energy change for E–S binding is given by $\Delta G_s = -RT \ln (1/K_s)$. Finally, the entropy change for E–S binding can be estimated from $\Delta S_s = (\Delta H - \Delta G)/T$. The preceding analysis is useful when comparing the

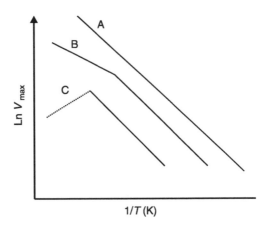

**Figure 11.6** Arrhenius plot for a reaction catalyzed by an enzyme which (A) is stable, (B) undergoes a conformational change, and (C) denatures at a high temperature.

efficiency of enzyme binding to several substrates. The Arrhenius plot for a reaction catalyzed by an enzyme is shown in Figure 11.6.

## 11.4   ENZYME INACTIVATION

### 11.4.1   ENZYME STRUCTURE

With the exception of some catalytic RNA molecules all enzymes are proteins. Enzymes are composed of amino acids but a significant number also contain carbohydrate. Glyco-proteins tend to be more soluble and stable as compared to carbohydrate-free enzymes. Many food enzymes also appear as multiple forms that differ largely in their carbohydrate content. The presence of one or more metal ions (Ca, Zn, Mn) can also increase enzyme stability.

Some enzymes are **complex globular proteins** that possess co-enzymes or nonprotein components. For instance, the heme proteins (e.g., myoglobin, peroxidase, catalase) contain an apo-protein component bound to heme. The hierarchy of enzyme primary, secondary, and tertiary structure is as described for other food proteins in Chapter 6. **Multi-subunit enzymes** have identical subunits each with its own active site. Alternatively, a **catalytic subunit** with an active site is associated with a **regulatory subunit** that allows the enzyme to respond to low molecular weight, allosteric, modulators. Lactose synthese, which catalyzes the synthesis of lactose, has a catalytic subunit associated with a regulatory subunit called $\alpha$-lactalbumin. This regulatory unit contains one molecule of tightly bound $Ca^{2+}$. Removal of the $Ca^{2+}$ ion using the chelator EDTA causes $\alpha$-lactalbumin to form a molten globule state, a kind of denatured protein structure, at room temperature. Other metallo-proteases contain $Zn^{2+}$ ions that are intimately involved in the catalytic process.

### 11.4.2   HIGH TEMPERATURE DEACTIVATION KINETICS

Loss of enzyme activity is called inactivation or deactivation. Enzymes that catalyze food deterioration must be inactivated in order to preserve food quality and extend shelf-life. Other research is focussing on ways to stabilize enzymes against inactivation in order to extend their application in industry.

The inactivation of most enzymes follows a one-step process (Eq. 11.23). The inactivated enzyme state (I) is formed by a number of chemical and physical process such as protease attack, adsorption to solid surfaces, and protein aggregation. Inactivation can also occur by way of two **consecutive** (Eq. 11.24a) or **parallel** (Eq. 11.24b) reactions.

$$N \xrightarrow{k} I, \tag{11.23}$$

$$N \xrightarrow{k'} N^* \xrightarrow{k''} I, \tag{11.24a}$$

$$N \xrightarrow{k'} I_1, \qquad N \xrightarrow{k''} I_2. \tag{11.24b}$$

Irrevesible inactivation (Eqs. 11.23, 11.24a, and 11.24b) follows first-order kinetics. To examine deactivation kinetics heat a sample of enzyme (or fresh food component containing some enzyme) using a thin-walled test-tube immersed in thermostatic water-bath. After a known time interval ($t$) remove the enzyme sample from the heat source, cool, and determine the remaining enzyme (residual) activity.

Suppose that a food sample is subjected to a well known temperature–time heating regime. Following thermal treatment, the sample is cooled to room temperature and the residual enzyme activity ($N$) is determined. In this hypothetical study, $N$ decreases with heating time ($t$) in accordance with a first-order relation (Eq. 11.25a). For enzymes such as peroxidase, inactivation proceeds via consecutive or parallel process as described by Eq. 11.25b:

$$[N] = [N]_0 \exp(-kt) \tag{11.25a}$$

$$[N] = [N]_0 \exp[-(k' + k'')t] \tag{11.25b}$$

where $[N]_0$ is the starting enzyme activity before heat treatment. The rate constant for deactivation can be determined from the semi-log equation

$$\ln[N] = \ln[N]_0 - kt. \tag{11.26}$$

Therefore, a graph of $\ln[N]$ versus time ($t$) gives a straight line with a gradient $k$. For a consecutive or parallel reaction, a semi-log plot produces two straight-line segments with two gradients equal to $k'$ and $k''$.

From elementary kinetics, two **indices of enzyme stability** can be defined (Eq. 11.27a). The half-life ($t_{1/2}$) for an enzyme is the time required for the intial activity to decline by 50%. The second stability index is the decimal reduction time ($D$-value) which is the time needed for enzyme actvity to decline by 90%. The temperature dependence of enzyme stability is described by Eq. 11.27b.

$$t_{1/2} = \frac{0.693}{k} \quad \text{and} \quad D - \text{value} = \frac{2.3}{k} \tag{11.27a}$$

$$Z = \frac{T_1 - T_{\text{ref}}}{\log D_{\text{ref}} - \log D_1} \tag{11.27b}$$

Thus a plot of $\log D$ versus temperature ($T$) gives a straight line. The $Z$-value is the temperature interval necessary to produce a 10-fold change in the $D$-value. The $Z$-value can be estimated from $D$-values determined at two temperatures ($T$ and $T_{\text{ref}}$). Figure 11.7 is a semi-log graph showing variations of $D$-value with temperature.

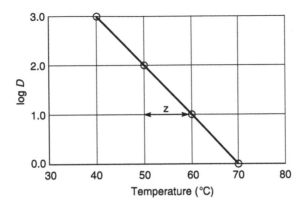

**Figure 11.7** A semi-log graph showing variations of $D$-value with temperature. The $Z$-value ($^{\circ}$C) is the temperature interval leading to a 1-log change in the $D$-value.

### 11.4.3 High Pressure Deactivation

Reactions that proceed with a net increase in volume ($\Delta V$ = positive) will be retarded by ultra-high pressure (UHP) treatment. In contrast, reactions accompanied by a net volume decrease ($\Delta V$ = negative) increase during UHP processing. Enzyme deactivation by UHP treatment usually follows first-order kinetics (Eqs. 11.28a–c):

$$\text{reactant (N)} \xrightarrow[\Delta P]{k} \text{Product (I)}, \tag{11.28a}$$

$$[\text{N}] = [\text{N}]_0 \, \exp(-kt), \tag{11.28b}$$

$$\Delta V = V_{\text{product}} - V_{\text{reactant}}. \tag{11.28c}$$

There is an important analogy between UHP inactivation and thermal inactivation of enzymes. With increasing pressure the rate of enzyme deactivation increases in accordance with the **activation volume change** ($\Delta V^{\#}$): $\ln k = \ln A + (\Delta V^{\#})P$. Different enzymes respond differently to UHP processing due to variations in their adiabatic compressibility (Chapter 7, Section 7.4.3). The application of UHP to the control of food enzymes is described in Section 11.6.5.

### 11.4.4 Enzyme Indicators for Meat Processing

Heat treatment is necessary to ensure that **meat** is free from **food-borne diseases** and microbial pathogens.[8] There is a significant risk of disease transfer when meat is imported from countries where foot and mouth disease is endemic. Other meat borne **viral agents** are also of concern, including rinderpest, fowl pest, Newcastle disease, and African swine fever. Meat is also a potential source of bacterial food borne pathogens, e.g., *Salmonella* and *Escherichia coli*. In response to safety concerns the USDA-FSIS (1985) recommended that meats should be cooked to certain **end point temperatures** (EPTs) before importation (Table 11.5). To prevent disease transfer the centermost part of the imported meats should have reached a certain EPT during cooking (Table 11.6).

Several tests for meat thermal history were proposed. To perform the **pink-juice test**, a cube of certified dimensions is removed from the center of a meat sample. The cube is compressed and any juice exuded is examined. The presence of pink coloration, showing undenatured myoglobin, shows inadequate heating. An alternative test for EPT is called the *protein*

**Table 11.5**

**Suggested end point temperatures (EPTs) for some meat products**

| Meat product | EPT (°C) |
|---|---|
| Cooked beef, roast, corned beef | 62.8 (145°F) |
| Cured/smoked poultry products | 68.3 (155°F) |
| Imported canned ham | 69 (156°F) |
| Poultry roll and other | 71.1 (160°F) |

**Table 11.6**

**Thermal inactivation parameters for some food pathogens and enzyme indicators of EPTs**

| Bacterium/enzyme | Temperature (°C) | D-value (min) | Z-value (°C) |
|---|---|---|---|
| *Escherichia coli* | 65 | 1.7 | 6.0 |
| *S. senftenburg* | 65 | 3.4 | 5.6 |
| Lactate dehydrogenase | 64 | 449.4 | 3.8 |
| Lactate dehydrogenase | 70 | 12.2 | |
| Acid phosphatase | 71.1 | 12.2 | 6.3 |
| Creatine kinase | 64 | 0.7 | 4.8 |

**coagulation test**. A sample of meat is homogenized with water and filtered. The concentration of soluble protein is then determined using the biuret assay. High levels of soluble proteins, compared to a control sample, is an indication of a failed test. These EPT tests are subjective and prone to error. For example, the judgment "pink" can vary with the prevailing lighting conditions. Meat protein coagulation is affected by sample pH as well as the sample thermal history (Chapter 12, Section 12.7.3).

Recently, more reliable and sensitive tests for meat thermal history were developed using **enzymatic** techniques. Enzymic EPT indicators are enzymes whose levels remain constant during meat storage. However, these enzymes show rapid deactivation when meat is heated above the USDA–FSIS recommended EPT. Suitable enzyme indicators are also more heat stable than some common meat-borne viral and bacterial pathogens.

An early enzyme-based EPT indicator is the **APIZYM**® system.[9] The enzyme test kit contains 19 ampoules of dehydrated substrates for testing 19 different enzymes. To use this test kit, the analyst adds a potential enzyme source to each ampoule and examines this for a color change 1–4 h later. Using APIZYM, it was demonstrated that raw beef had high activities for nine enzymes. Mild heat treatment led to the loss of most of these enzymes with the exception of **leucine amino peptidase** (LAP) and **acid phosphatase** (AP). These two enzymes were therefore suggested as potential EPT indicators for meat samples heated to 62–64°C. Another enzymatic EPT indicator is **glutamate oxaloacetate transaminase** (GOT) found in most meat. The range of GOT activity found in raw chicken homogenate was 7,700–18,000 U/ml. The average GOT activity decreased by 1000-fold within 3 min of heating chicken samples at 73–75°C. Roasted chicken drumstick, breast, thighs, and legs had $11.5 \pm 1.9$ Units of GOT per ml. Since commercial test kits for GOT are available it is possible that this test could be adopted for checking whether chicken has been adequately heated (Table 11.5).

Perhaps the most promising enzyme EPT indicator for meat is **lactate dehydrogenase** (LDH).[10] There are four LDH isoenzymes found in vertebrates. Each isoenzyme is built from two proteins subunits (H and M). Heart muscle contains the $H_4$ isoenzyme, which comprises 4-H subunits. In meat, over 90% of the total LDH activity consists of the $M_4$ isoenzyme. LDH activity decreases when meat was heated to a temperature of 63–66 °C. Inactivation of LDH present in meat was higher than the extent of inactivation for meat homogenates. The LDH system satisfies one of the key requirements for a potential enzymatic EPT indicator. This is the need for a simple and affordable assay. Commercial assays for LDH have been around for a long time. LDH catalyzes the reduction of pyruvate by nicotinamide adenine dinucleotide (NADH):

$$\text{pyruvate} + \text{NADH} \rightarrow \text{lactate} + \text{NAD}. \tag{11.29}$$

The decrease in NADH concentration can be followed by the decrease in UV absorbance at 340 nm. Alternatively, pyruvate can be reacted with 2,4-dinitrophenylhydrazine to form a blue product monitored at 460 nm.

Table 11.6 shows some typical $D$- and $Z$-values for some meat enzymes. Thermal deactivation values for some common bacterial pathogens are also shown for comparison. For instance the $D$-value for the pathogen *E. coli* Oi57:H7 is 1.7 min compared to 12.2 min for AP and approximately 449.4 min for LDH. This suggests that a meat sample heated to destroy LDH or AP activity is likely to be free from *E. coli O157*. Creatine kinase, on the other hand, is relatively unstable and not a good EPT indicator.

## 11.4.5 ENZYME INDICATORS FOR VEGETABLE BLANCHING

The action of endogenous enzymes affects the sensory, nutritional quality, and shelf-life of fresh vegetables. Blanching technology arose in the 1930s as a way for extending the shelf-life of fresh produce. Mild heat is applied thereby inactivating enzymes and preventing undesirable changes in fresh-food quality. The temperature for blanching is generally low enough to retain "fresh-like" quality.[11]

The role of blanching is to inactivate endogenous enzymes and thereby stabilize the sensory characteristics of a product. As well as causing enzyme deactivation, blanching also produces some sterilization and a reduction of bacterial numbers. Exposure to high temperatures encourages the removal of gasses from within vegetable tissues and induces wilting and shrinkage which helps with subsequent packaging. Finally, blanching can facilitate peeling where necessary. Blanching is applied to vegetables before canning, pickling or freezing. In contrast, fresh, minimally processed or dried vegetables are not usually pre-blanched. The process of blanching involves exposure to steam or hot water (75–100°C) for various prescribed times.

The disappearance of certain enzyme activities during blanching correlates with extended shelf-life. According to Jocelyn blanching enzyme indicators can be identified as follows:

(1) Blanch a product using various temperature–time combinations.
(2) Freeze the pre-blanched product and store for known times.
(3) Thaw and evaluate the quality, preferably using a sensory panel.
(4) Test the product for a range of endogenous enzyme activity.
(5) Identify the correlation between enzyme activity measurements and sensory scores.

From such studies, the loss of **peroxidase** activity *was shown to be the* best indicator of quality (flavor) retention in blanched and frozen vegetables.[12]

Arguments promoting peroxidase as a blanching indicator enzyme include:

(1) Its widespread distribution. Peroxidase is found in most plant species.
(2) High heat resistance. Peroxidase is arguably the most heat resistant enzyme within plant tissue. Loss of peroxidase activity during thermal processing is a sure sign that all other enzymes have been inactivated.

(3) Wide availability of simple assays. Many simple and highly sensitive peroxidase assays have been developed for clinical use and can be adapted for blanching studies.

However, there appears to be no direct link between peroxidase activity and loss of quality in frozen vegetables. Though recent results suggest that peroxidase may catalyze enzymatic browning and lignification, these reactions do not appear to lead to major defects in stored vegetables. Second, denatured peroxidase has a tendency to regenerate activity during storage. Peroxidase **regeneration** adds to the heat resistance of this enzyme. There is a possibility of over-blanching when the applied process time is sufficient to denature peroxidase. Third, peroxidase occurs as multiple forms. The heat stability and distribution of the various peroxidase isoenzymes within different plant tissues is not necessarily uniform. Finally, about 1–10% of the peroxidase activity associated with plant tissue is not sensitive to heat.

There were large variations in the peroxidase activity from nine common vegetables (Table 11.7). The highest and lowest peroxidase activities were found in cabbage (562 U/kg) and onion (8 U/kg).[13] Blanching 5–10 mm pieces of vegetables at 75–95°C produced decreases in peroxidase activity, although the rate of inactivation was different for different foods. The final levels of peroxidase activity depended the blanching temperature, size of vegetable pieces, nature of the vegetable, initial peroxidase activity present, and the total starch content. The rate of peroxidase inactivation in leafy vegetable was faster, probably because of faster heat transfers. On the other hand, the gelatinization of high starch samples (e.g., potato) may have hindered heat transfer. It was concluded that complete inactivation of peroxidase was not essential for quality retention in blanched vegetables.

Williams and co-workers[14] describe an approach for identifying actual endogenous enzyme(s) responsible for quality loss in blanched and frozen vegetables. The major flavor defect enzyme was identified by (1) blanching a food product to inactivate all endogenous enzymes, (2) adding several purified enzymes to food samples, freezing and storing for prescribed times, and (3) thawing and accessing product quality. This study showed a clear correlation between flavor defect detected by sensory analysis and the addition of **lipoxygenase**. The wide distribution, relatively low heat resistance and direct involvement in off-flavor formation are cited as good reasons for selecting lipoxygase as the blanching indicator. By contrast, there are few rapid assays for monitoring lipoxygenase levels in crude plant tissue extracts. A simple assay using lipoxygenase is the measurement of absorbance changes at 243 nm due to the oxidation of linoleic acid to a conjugated diene. This UV-based method is not applicable for highly colored samples. The starch–iodide method provides a quantitative colorimetric assay for lipoxygenase. The detection is based on the oxidation of colorless iodide

**Table 11.7**
**Peroxidase activity in some common vegetables**

| Plant | Common name | Peroxidase activity (Units/kg) |
|---|---|---|
| Brassica oleracea | Cabbage | 562 |
| Phaseolus vulgaris | Green bean | 515 |
| Curcubita pepo | Squash | 361 |
| Spinicia oleracea | Spinach | 195 |
| Apium graveolens | Celery | 41 |
| Solanum tuberosum | Potatoes | 40 |
| Allium porrum | Leeks | 37 |
| Dacus carata | Carrots | 12 |
| Allium cepa | Onion | 8 |

Adapted from Muftugil.[13]

ions by linoleic acid hydroperoxide to form free iodine, which then forms a blue–black complex with soluble starch.

## 11.5 ENZYME ISOLATION AND PURIFICATION

Food enzymes have to be isolated in various states of purity before being fully characterized. Studies of partially purified food enzymes with respect to their inhibition characteristics, substrate preference, and activators, for example, can lead to improved understanding of their control. Some understanding of enzyme isolation or downstream processing is also important for industrial enzymes, which are used as processing aids in the food industry.

### 11.5.1 RAW MATERIAL FOR ENZYME EXTRACTION

The raw material for enzyme extraction is determined by an analyst's particular interests. Polyphenol oxidase (PPO), which catalyzes enzymatic browning, is ubiquitous in plants. The focus of any given study is usually to understand the effect of PPO on the quality of specific commodities. Grapes, bananas, pears, and apples might be of interest as a source of enzymes in those regions where these particular commodities have significant economic importance. In France, researchers of PPO used pears and grapes as the enzyme source. By contrast, much classical work on PPO involved the enzyme obtained from potato. Lipoxygenase (LOX) from soybean has been widely studied because this enzyme is responsible for the formation of "beany" flavor. LOX from fish muscle was recently studied owing to its possible involvement in the formation of off-flavor and loss of texture. It is further necessary to choose an appropriate tissue or organ for enzyme extraction. Plant enzymes are likely to be extracted from the vegetative parts (fruits, leaves or tubers) or seeds with economic interest. It is rare to isolate enzymes from the whole plant. The selection of animal tissue is also important, because enzymes from different tissues and organs have distinct characteristics.

Material purchased from "a local market" is mostly unsuitable for enzyme research. Market purchases have an uncertain origin. The age at harvest, temperature history, and details of growing or agronomic practices will all be uncertain. With the advent of modified atmosphere storage, fruits can be stored for long periods, without apparent adverse effects on their appearance. A lack of freshness adversely affects the level of enzymes in plant tissue. In the case of muscle, a degree of ageing is desirable before enzyme extraction. Relatively fresh muscle is usually in a pre-rigor state where high levels of ATP act to maintain actin and myosin in a dissociated state. The soluble actin and myosin can later precipitate as the level of muscle ATP falls, thereby trapping and removing a desired enzyme. Allowing the formation of acto-myosin before enzyme extraction avoids excessive enzyme loss.

It is generally advisable to use a reputable supplier of raw materials for enzyme extraction. Another option is to use plant material grown by the investigator or a close associate under well defined conditions of temperature, light, and fertilizer treatment. Universities and research institutes frequently maintain research farms for the purpose of providing reliable research material.

### 11.5.2 TISSUE DISINTEGRATION AND ENZYME EXTRACTION

The Warring blender is a common method for disrupting plant or animal tissue for enzyme extraction. The construction of this apparatus is virtually identical to the domestic food blender. Other common methods for food disintegration include the use of **high energy ultrasound**, digestion with lysozyme or treatment with an **organic solvent**, usually toluene. With mechanical disruption (blending, ultrasonic treatment) it is necessary to avoid excessive sample heating which may lead to enzyme deactivation. Using a pre-chilled sample with intermittent blending or ultrasonication allows time for heat dissipation. Samples can also be cooled using an ice-bath. A buffer solution is usually added to the food sample before its disintegration and enzyme extraction. A weight ratio ranging from 1:1 to 1:4, food sample to added buffer, is normal so long as the sample viscosity is not too great. Nonmechanical methods for tissue and cell disruption include treatment with lysozyme (Figure 11.8)

Plant tissue extracts prepared as described so far will soon begin to darken due to **polyphenol oxidase** (PPO) mediated oxidation of phenol compounds originating from the plant vacuole. Enzymatic browning leads to brown pigments which are amphipathic and able to bind proteins via hydrogen bonding and hydrophobic interactions. The polymerization reaction itself involves the formation of a quinone intermediate that will also react covalently with proteins. Browning can be prevented by adding antioxidants to the plant tissue *before* the tissue disruption stage. Addition of up to 30 mM ascorbic acid serves as an adequate inhibitor for enzymatic browning. Sulfur dioxide, an efficient inhibitor of browning, also disrupts protein disulfide bonds and is therefore not suitable for enzyme studies. An alternative strategy for avoiding browning is to add polyvinyl pyrrolidone (PVP) to bind polyphenols. Another reason for the loss of enzyme activity during isolation is the action of proteases present in plant or animal extracts. Commercial mixtures of protease inhibitors are available which inhibit a wide range of protease types. Methods of cell breakage are shown in Figure 11.8.

### 11.5.3   ISOLATING MEMBRANE BOUND ENZYMES

Membrane associated enzymes or bound enzymes are enzymes which are enclosed within cell organelles or adsorbed on cell debris during the tissue disruption phase (Section 11.5.2). To avoid loss of bound enzymes during the filtration and clarification phases, this enzyme fraction should be solubilized with the aid of a surfactant. Addition of low (0.1–0.3% w/w) concentrations of non-ionic surfactants such as Tween 20 or Triton-X100 in the enzyme extraction buffer will help to solubilize the "bound" enzyme fraction. Indeed, a simple way to detect the presence of bound and free enzyme fractions is to assay for enzyme activity with and without added surfactant. Apart from the requirement for a non-ionic surfactant, the isolation procedure for bound enzymes is no different from methods described for soluble enzymes.

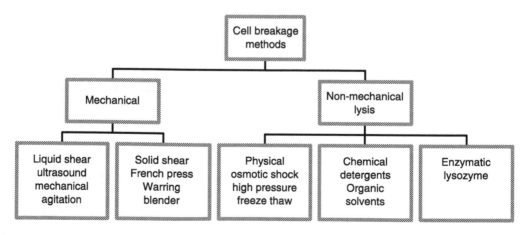

**Figure 11.8**   Methods of cell breakage.

### 11.5.4   MICROBIAL ENZYMES

A great many enzymes for food processing are derived from microbial sources. The majority are extra-cellular enzymes, which are secreted into the external growth medium during the growth of microorganisms in large bioreactors or fermentors. The fermentation systems are of two types: **semi-solid fermentation** and **submerged culture**. Most fungi are grown on semi-solid food residues (wheat bran + nutrients) in rotating drums to provide aeration and mixing. The majority of *Bacillus* sp. are grown in submerged culture using liquid broth (mixture of corn starch, soybean meal, casein, sodium phosphate buffer salt) in temperature-controlled batch fermentors (large vessels) fitted with ports to allow aeration and mixing, and with probes for monitoring pH and oxygen levels. Aseptic conditions are maintained during enzyme production so no adventitious microorganism, other than the enzyme-producing species, grows.

### 11.5.5   CLARIFICATION OF THE CRUDE ENZYME EXTRACT

Newly prepared enzyme extracts contain suspensions of particulates. Homogenizing plant tissue will yield cellulosic fiber, fat, starch, and insoluble proteins, as well as low molecular weight phytochemicals. After homogenization, animal tissue will yield muscle fragments, insoluble structural protein, and fat. Such a suspension needs to be clarified before further processing. Microbe fermentation will also contain a suspension of colloids, solid media components, bacteria cells or fragments of fungal mycelium, as appropriate.

Filtration is a simple way to achieve clarification. For laboratory scale preparation of food enzymes the filter may range from paper (Whatman) to glass fiber. Industrial scale preparation employs drum filters or leaf filters. Precipitants (calcium chloride, polyelectrolytes) or diatomaceous earth may be added to facilitate filtration. Centrifugation is suitable for clarifying samples of several liters. During filtration, enzyme loss may occur due to protein adsorption on the filter material. Using a buffer with an adequately high ionic strength (0.1–0.2 M) can reduce the loss of enzyme activity via this route.

### 11.5.6   ENZYME CONCENTRATION AND RECOVERY

Common methods for concentrating industrial enzymes include vacuum evaporation (35–50°C) and ultrafiltration (5°C). The low temperature conditions usually result in high recovery of enzyme activity during ultrafiltraton. However, consideration should be given to the possibility of membrane fouling and protein adsorption to membranes. The enzyme concentrate from ultrafiltration is usually cloudy and may need further filtration to clarify it. The final stage is to add diluents or "extenders" as well as stabilizers (salts, starch hydrolysates, polyols) before sale. Antioxidant and antimicrobial agents (e.g., sodium benzoate, parabens) may be added provided that they are not toxic. Solid commercial enzymes are produced by low temperature spray drying of enzyme solutions previously pre-concentrated by ultrafiltration. Enzymes may be partially purified to remove nonenzymatic protein and thereby increase the specific activity. The most common methods for purification are salt (ammonium sulfate) fractionation or solvent (acetone, ethanol) precipitation.

The preceding methods can also applied to concentrate enzymes on the small scale. Further fractionation may be necessary before fundamental studies can begin. Chromatographic methods are used for enzyme purification. On a laboratory scale there are many methods of protein purification. Most purification methods involve chromatography using affinity, ion exchange, or gel-permeation supports. Other useful techniques for enzyme isolation include isoelectric focusing and two-phase liquid–liquid extraction.

## 11.6   FOOD QUALITY RELATED ENZYMES

Living food materials contain a vast number of **endogenous enzymes**. The activity of such enzymes produce undesirable, as well as some desirable, changes in the quality of fresh foods. Deteriorative changes occurring within processed foods are largely chemical (Chapters 9 and 10). Processed foods that become contaminated by microorganisms will undergo biological spoilage. Hydrolases are responsible for the degradation of biopolymers (proteins, poly-saccharides—lipids may be considered "honorary" biopolymers) thereby bringing about changes in the feel and texture of raw foods.

### 11.6.1   PLANT CELL WALL ENZYMES AND FOOD TEXTURE

From a food perspective, plant cell walls provides texture and dietary fiber. The biological function of the cell wall is (1) to provide tensile strength and mechanical support to cells and thereby prevent cell rapture, (2) to protect cells from excessive water loss, insects, and pathogen attack, (3) to allow cell growth, expansion, and increase in size, and (4) to allow intracellular communication between adjoining cells. There are quantitative as well as qualitative changes in cell wall material during cell division, enlargement, maturation, and senescence. Common experience shows that plant materials grow tougher and woody with age. The increase in texture is the result of **lignification** and the formation of *tyrosine* **cross-links** between cell wall components. Fruits also soften during ripening owing to attack by fungi. The loss of texture is due to the breakdown of plant cell wall components such as **pectin**, **cellulose**, **hemicelluse,** and a **protein** called **extensin**. Therefore, changes in the state of plant cell walls affect food quality.

The structure and composition of the plant cell wall is not fully understood. So there is no generally accepted model for the plant cell wall[e]. Moving outwards from the plant cell interior, we first encounter a phospholipid bilayer comparable to the *cell* **membrane** that surrounds all animal cells. Beyond the lipid membrane, plant cells have an additional three layers:

(1) *The* **primary cell wall**. This develops from the fusion of small vesicles floating on the cell membrane immediately after cell division. The primary cell wall consists of cellulose microfibrils embedded within **matrix polysaccharides**. This gel-like matrix phase is composed of pectin, hemicellulose, and glycoproteins.

(2) **Middle lamella**. This layer which is also rich in pectin is shared by adjacent cells and serves as the outer covering of plant cells during growth and enlargement.

(3) **Secondary wall**. This, the innermost cell layer, is highly rigid and comprises cellulose, hemicellulose, and lignin. After plant cell growth the secondary wall forms in the form of encrustations deposited on the surface of the middle lamella. It is the thick secondary cell wall that produces a tough woody texture of aging plant tissue[f]. Because of its impermeable layer, the formation of a woody secondary cell wall is usually followed by cell death.

The multitude of cell wall components act as substrates for **endogenous enzymes** including, **peroxidase**, **cellulases**, and **pectinases**. Peroxidases act on a number of phenolic substrates, generating free radical precursors required for lignification. However, more subtle effects on plant food texture have been suggested by research at the Institute of Food Research (UK) led by Keith Waldron. According to this work, peroxidase mediated cross-linking of

---

[e]The four major models of plant wall architecture (the covalently cross-linked model, tether model, diffuse layer model or stratified model) will be found at the University of Georgia (US) Complex Carbohydrate Research Center website, http://www.ccrc.uga.edu/~mao/intro/ouline.htm (accessed January 2004).

[f]The chemical composition of plant cell walls is described at the University of Hamburg website, http://www.biologie.uni-hamburg.de/b-online/e00/index.htm (accessed January 2004).

hydroxycinnamic acids and related phenols (Figure 11.9–11) may explain why some plant foods retain their crispy texture after heating. Chinese watercress is one of a number of plants that remain crispy after cooking. In the model offered by Waldron, biopolymers involved in cell–cell adhesion are reinforced by tyrosine or phenolic cross-links. Attempts are being made to link ferulic acid and other peroxidase substrates with the response of plant tissues to thermal processing.[15]

The structure of pectin was described in Chapter 4. Pectin functions as an intracellular cement between cells. Degradation of pectin involves two enzymes, **pectin methyl esterase** (PME) and **polygalacturonic hydrolase** (PG). First, PME hydrolyses the methanol ester bond leading to free galacturonic acid residues. The acidic pectin polymer is now hydrolyzed by PGase resulting in the depolymerization of pectin. **Cellulases** may also be involved in the degradation and weakening of plant cell wall polymers. Recently, the role of PME and PGase in the control of fruit texture was called into doubt. Genetically modified tomatoes with reduced PME and PGase levels undergo softening during ripening. Enzymes that hydrolyze other plant cell wall polysaccharides such as **cellulases and xylanases** may have a bigger role in controlling texture loss in intact fruits or vegetables.

Pre-heating vegetables before canning is believed to improve the quality of the final product. Similarly, potatoes are subjected to preheating treatment before cooking and roller drying. Mild heat treatment may allow pectin de-esterification by PME. This, together with the release of $Ca^{2+}$ from intracellular stores, leads to pectin cross-linking and improved texture. In the case of potato, pre-heating appears to encourage cell separation rather than disruption and so avoids a pasty-tasting product.

The **expansins** are cell wall loosening proteins which are thought to disrupt H-bonding interactions between cellulose microfibrils and matrix polysaccharides[g]. Expansin genes are expressed during fruit ripening and expansins are thought to play a role in pollination, although their exact function remains a mystery. The amino acid sequence for $\alpha$-expansin and $\beta$-expansin have been deduced from their gene structures. The $\sim$25 kDa proteins have two domains. There is a cysteine-rich domain, which resembles an endoglucanase (carbohydrate degrading enzymes). The second domain is at the tryptophan-rich carboxyl terminal with a possible role in polysaccharide binding. No enzymatic activity has been detected for $\alpha$-expansin or $\beta$-expansin. The effect of expansin activity for fresh food texture remains to be determined.[16]

## 11.6.2 ENZYMIC BROWNING BY POLYPHENOL OXIDASE

The discoloration of many fruits, vegetables, and mushrooms is due to a reaction catalyzed by polyphenol oxidase (PPO), also called tyrosinase.[17] PPO is a copper-containing enzyme with **catecholase** and **cresolase** activity. In the former case PPO is able to remove hydrogens from catechol to form diquinone. Cresolase activity refers to the hydroxylation of cresol. Most of the research on PPO was done with mushroom tyrosinase (molecular weight 128 kDa), which contains four subunits. Each subunit has one mole of copper per molecule and two binding sites for aromatic compounds including phenolic substrates.

$$\text{Cresolase} : \text{monophenol} + \text{oxygen} \rightarrow \text{dihydroxyphenol} + \text{water}.$$
$$\text{Catecholase} : \text{diphenol} + \text{oxygen} \rightarrow \text{diquinone}.$$
$$\text{Nonenzymic browning} : \text{quinone} + \text{protein} \rightarrow \text{brown pigments}. \tag{11.30}$$

---

[g]The reader is referred to the expansin web page of Professor Daniel Cosgrove, Pennsylvania State University, Department of Biology, http://www.bio.psu.edu/expansins

The structures of some plant PPOs are beginning to be elucidated using molecular biology techniques. Most appear to be quite different from mushroom tyrosinase. The DNA sequence for PPO from beans (*Vici faba*) corresponds to a 58 kDa protein. The enzyme protein appears to have a conserved sequence for two copper-binding domains which matches the 59 kDa tomato PPO, hemocyanins, and tyrosinases from a wide range of sources.[18] The cDNA sequence for spinach PPO encoded for a 64 kDa protein. The amino acid sequence at the N-terminus is thought to serve as a signal for transport of PPO into the chloroplast. The C-terminal has a large number of hydrophobic groups surrounded by basic amino acids, which suggests that these structures help to position PPO within the thylakoid lumen. Spinach PPO had 50% sequence homology with PPO from other souces. The central domain has a conserved motif, N-terminal to a presumptive Cu-A site that is not found in tyrosinases.[19]

Mushroom PPO has a **pH optimum** of 6.0–7.0. Plant cell PPOs show more varied behavior with optimum activity ranging from pH 5 to 8.0. The PPOs from sweet cherry and taro had their optimum activity at pH 4–4.5. In general, activity declines below pH 4. As described above, the catecholase activity of PPO uses $O_2$ to catalyze the dehydrogenation of catechols to orthoquinones. The cresolase activity results in the hydroxylation of phenols to catechols. PPOs have also been reported to possess peroxidase activity.[20] The **substate specificity** of PPO is very broad. A large number of para-substituted catechols are oxidized, including monohydroxy-, *o*-dihydroxyphenols and polyhydroxyphenols. The **natural substrates** for the browning reaction within plants include catechins, cinnamic acid esters, 3,4-dihydroxy-phenylalanine (DOPA) and tyrosine (Figures 11.9 to 11.12).[21] Many of these substrates are also acted upon by peroxidase when free hydrogen perxoide is provided. Compounds with a close resemblance to catechol function as competitive inhibitors for PPO (Figure 11.13).

A typical study showing the effect of PPO on fruit processing is illustrated by investigations of aubergine (eggplant) PPO. This vegetable is eaten fresh or dried. Browning of aubergine occurs during processing. Therefore the characteristics of aubergine PPO were determined by researchers from Turkey. Crude aubergine PPO showed an optimum browning

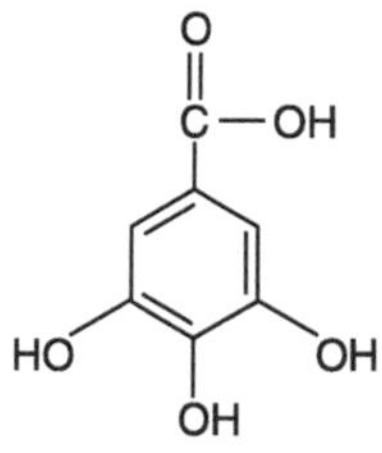

Gallic acid; 3, 4, 5-Trihydroxybenzoic acid

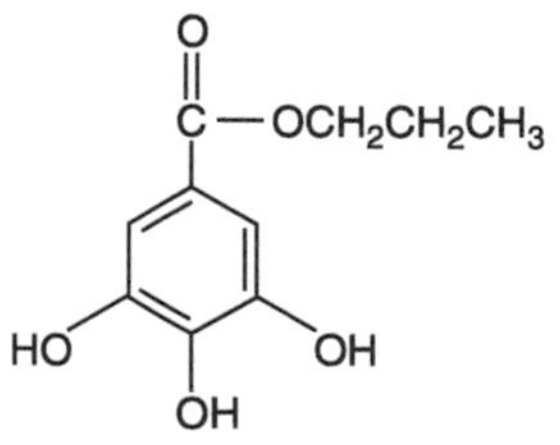

Propyl gallate

**Figure 11.9**   Simple phenolic susbstrates for peroxidase and polyphenol oxidase.

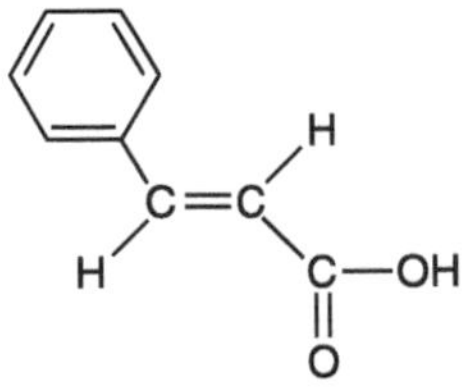

Cinnamic acid; *trans*-3-Phenylacrylic acid

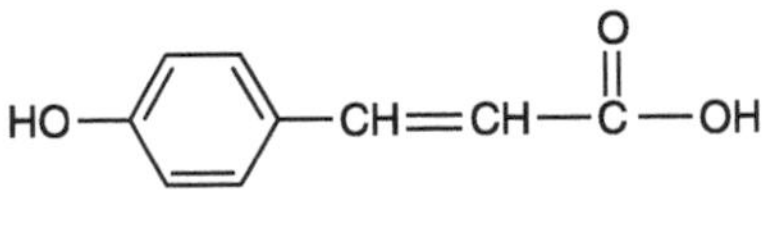

p-Caumaric acid; *trans*-4-hydroxy-cinnamic acid

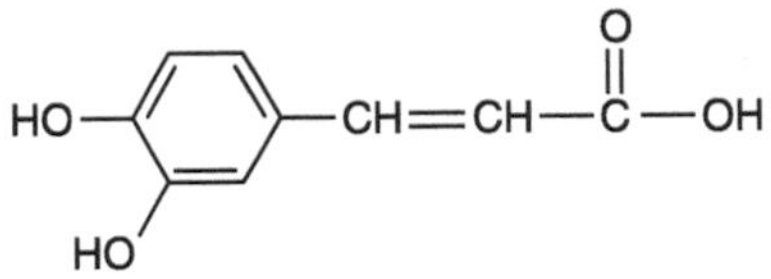

Caffeic acid; 3, 4-Dihydroxycinnamiccid

**Figure 11.10**   Cinnamic acid derivatives are natural substrates for peroxidase and polyphenol oxidase in fruits and vegetables.

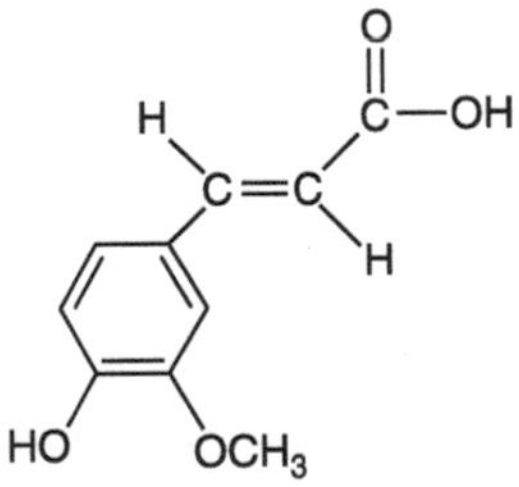

Ferulic acid (chlorogenic acid)

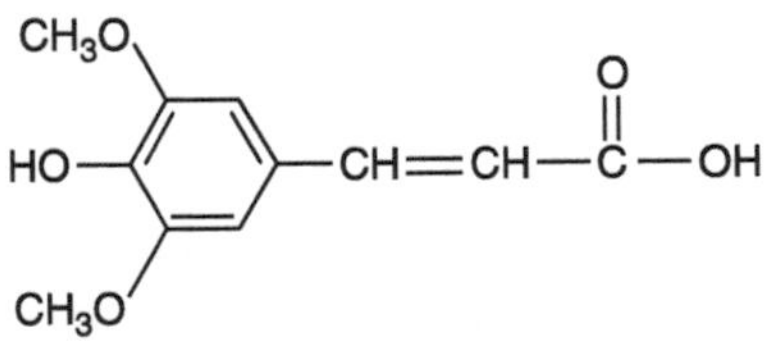

Sinapic acid; 4-Hydroxy-3, 5-dimethoxy-cinnamic acid

**Figure 11.11**   Some natural substrates for peroxidase and polyphenol oxidase.

activity at pH 7 with catechol as the substrate, and an optimum pH of 6 with 4-methylcatechol as the substrate. Like other plant PPOs, the aubergine enzyme was relatively easy to inactivate at temperatures above 40°C; the optimum temperature for activity was 20–30°C. Inhibitors for aubergine PPO included tropolone, D,L-dithiothreitol and glutathione.[22]

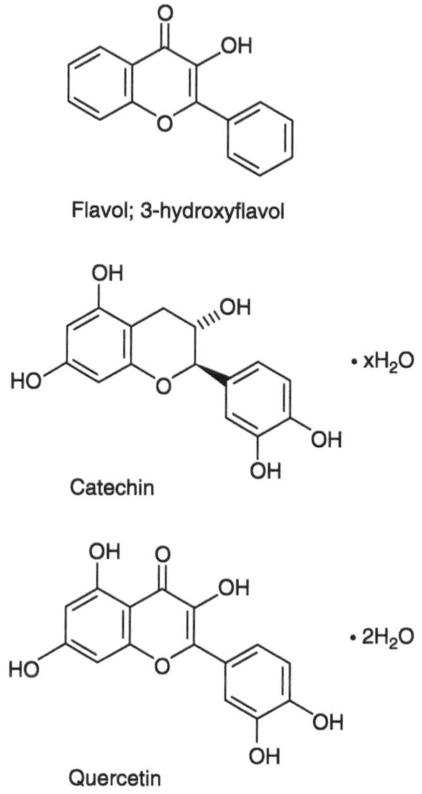

Flavol; 3-hydroxyflavol

Catechin

• xH$_2$O

Quercetin

• 2H$_2$O

**Figure 11.12**  Flavols and flavanone substrates for peroxidase and polyphenol oxidase.

### 11.6.3  LIPOXYGENASE: BLEACHING AND ODOR

Lipoxygenase (linoleate:oxygen oxidoreductases, LOXs) was discovered as the factor responsible for the **oxidation** of linseed oils. The **destruction of carotenoids** by soybean flour was ascribed to the presence of a heat susceptible agent, lipoxidase. It was not until the 1940s that Sumner and co-workers noted that lipoxidase, unsaturated fat oxidase, and carotene oxidase was the same enzyme.[23] LOX is a nonheme dioxygenase that catalyzes the transformation of polyunsaturated fatty acids to hydroperoxides. The substrates for LOX must contain a conjugated cis, cis-1,4-pentadiene-double bond. Soybean LOX was extensively studied in connection with the unpleasant **beany flavor** associated with soy products. The enzyme is also found in most plants (including algae), fungi (including yeasts), cyanobacteria, and mammals.[24] Animals contain 5-LOX whilst plants contain 9-LOX and 13-LOX. The numbers refer to the fatty acid carbon number at which the dioxygen molecule is introduced. A digramatic representation of the LOX reaction is shown in Figure 11.14.

Common **substrates** for LOX include linoleic acid, linolenic acid, and arachidonic acid. In plant cells, the hydroperoxide products from the LOX reaction are further transformed by one of two pathways: (1) the allene oxide synthase (AOS) pathway leads to traumatin (12-oxo-*trans*-10-dodecenoate) and jasmonic acid; and (2) the hydroperoxide lyase (HPL) pathway forms flavor molecules (hexenal, hexanols) as well as esters and other alcohols. In mammalian cells, the fatty acid hydroperoxide reacts to form bioactive products such as leukotrienes and lipoxin. These compounds are thought to be involved in **inflammation and pain generation**. The products from animal LOX lead to conditions like allergic rhinitis and asthma. A full discussion of the mechanisms of LOX reactions is beyond the scope of this text.

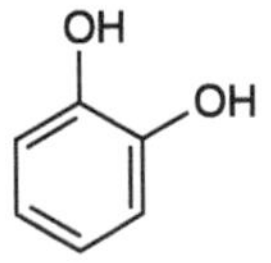

Catechol; 1, 2-Benzenediol; 1, 2-Dihydroxybenzene

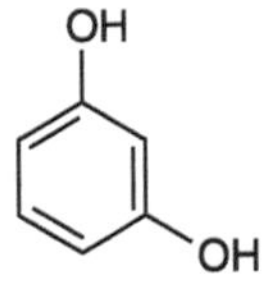

Rescorcinol; 1, 3-Benzenediol

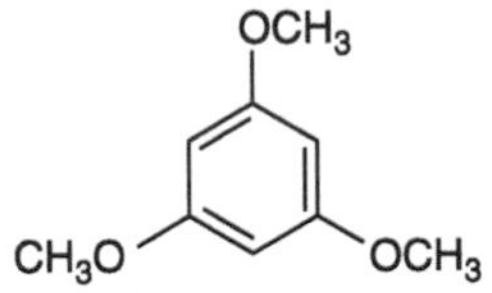

Phloroglucinol; 1, 3, 5 trihydroxy benzene

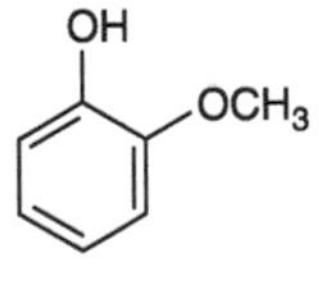

Guaiacol; Catechol monomethyl ether; 2-Methoxyphenol

**Figure 11.13** Comparing the structure of catechol (a polyphenol oxidase substrate) with some competitive inhibitors for this enzyme.

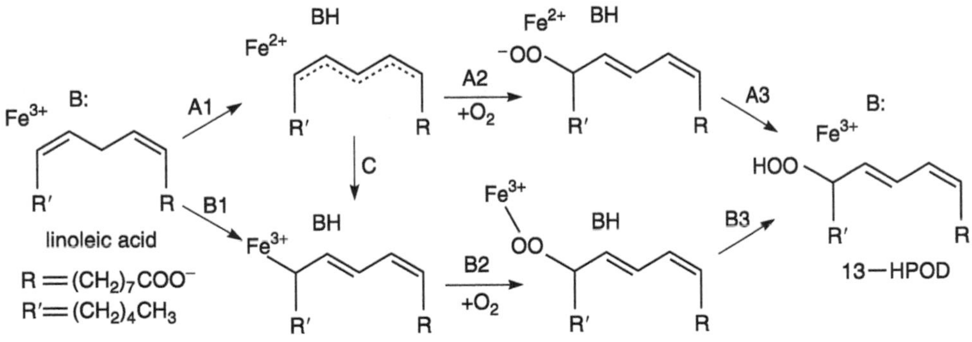

**Figure 11.14** A digrammatic representation of LOX action leading to hydroperoxide. The iron ($Fe^{2+}$) is located at the active site of LOX. The component A1 is a reducing agent. (From http://www.haverford.edu/chem/Scarrow/SLO/SoyLoxStr.html)

Interested readers are referred to an article by Whitaker, who describes the anaerobic and aerobic pathways for LOX.[24]

The effect of LOX on food quality is most readily apparent in relation to oxidative stability. LOX generated hydroperoxides and oxyradicals have an impact on flavor, texture, and color (see Figure 9.1). Until recently, the presence of a beany flavor was a major concern to soybean processors and producers of products such as soymilk and soy protein.

Using classical breeding and genetic engineering techniques, new soybean varieties were developed that have low LOX activity. LOX can also produce off-flavor in high fat fish such as mackerel. The inactivation of fat soluble vitamins by LOX has been reassessed recently from the standpoint of the loss of nutrients (undesirable). The carotenoid degrading activity is desirable when it is necessary to bleach wheat flour by the addition of soybean flour.[1]

### 11.6.4 PROTEASES

Meat tenderization is due to protein hydrolysis by **proteases** including the calcium activated calpains (at pH 7) or cathepsins (at low pH).[25] Hydrolysis of collagen by collagenases no doubt contributes to meat tenderization. The loss of meat texture is obviously desirable judging from the practice of hanging meat to mature.

### 11.6.5 CONTROL OF QUALITY RELATED ENZYMES

Ashie et al. classified methods for enzyme control as traditional and nontraditional.[26] Included in the traditional forms of enzyme control are (1) thermal treatments such as pasteurization, blanching, high temperature short time (HTST) treatment, and freezing; (2) control of water activity via dehydration or by salting; and (3) use of processing chemicals. Many chemical and processing aids affect enzyme activity and stability, including **chelating agents**, **reducing agents** (including **sulfites**), and **acidulants**. Many organic acids (e.g., oxalic acid, citric acid) can prevent PPO action by removing copper from the enzyme active site. Ascorbic acid and its derivatives function as inhibitors of browning owing to their **antioxidant** properties. Ascorbic acid and other reducing agents have two modes of action. First they can delete the oxygen needed for PPO activity. Second, the quinone intermediate formed by PPO action can be efficiently transformed back to a diphenol thereby preventing melanoidin formation. Sulfites are very effective in preventing browning, as they react directly with the quinone intermediate formed in the PPO reaction. A second mode for the sulfite effect on enzymes is via the reaction with disulfide bonds, leading to enzyme inactivation.

Some modern or nontraditional forms of control for endogenous food enzymes include ultrahigh pressure processing (UHP) and ionization. The former technology is relatively new and its scientific basis is not yet fully understood (Section 11.4.3). The type of enzyme, solvent conditions, and temperature affects the rate of enzyme deactivation under UHP conditions. The barostability of selected food enzymes could be ranked in the following order: peroxidase > PPO > catalase > phosphatase > lipase > pectin esterase > lactoperoxidase > lipoxygenase.

## APPENDIX 11.1   THE MICHAELIS–MENTEN EQUATION IN THE PRESENCE OF REVERSIBLE INHIBITORS

Deriving the Michaelis–Menten equations in the presence of reversible inhibitors follows the same principles described in Section 11.3.3.1 with minor changes. The enzyme mass balance equation (Eq. 11.9) with a CI present is now $E_T = E + ES + EI$. The specific rate of reaction is now written as before (Eq. 11.A1).

$$\frac{v_0}{[E_T]} = \frac{k_{CAT}[E][S]/K_s}{[E] + [E][I]/K_i + [E][S]/K_s}. \qquad (11.A1)$$

Rearranging 11.A1 to 11.A2 and 11.A3 follows elementary algebra.

$$\frac{v_0}{[E_T]} = \frac{k_{CAT}[E][S]/K_s}{[E](1 + [I]/K_i) + [E][S]/K_s},$$ (11.A2)

$$\frac{v_0}{[E_T]} = \frac{k_{CAT}[E][S]/K_s}{[E](1 + [I]/K_i]) + [E][S]/K_s}.$$ (11.A3)

As a final step we rearrange Eq. 11.A3 to the Michaelis–Menten equation (Eq. 11.18).

# References

1. Schwimmer, S., *Source Book for Food Enzymology*, The AVI Publishing Company Inc., Westport, CT, 1981.
2. Whitaker, J. R., *Principles of Enzymology for Food Science*, Marcel Dekker Inc., New York, 1994.
3. Wong, D. W. S., *Food Enzymes: Structure and Mechanism*, Chapman & Hall, London, 1995.
4. Stauffer, C. E., Enzyme assays for food scientists, in *Encyclopedia of Food Science and Technology*, Hui, Y. H., Ed., Academic Press, New York, 733–741.
5. Cornish-Bowden, A., The origins of enzymology, *The Biochemist* 19(2), 36–38, 1999.
6. Marmase, C., *Enzyme Kinetics: Physical Basis, Data Analysis and Uses*, Gordon & Breach, London, 1977.
7. Engels, P. C., *Enzyme Kinetics. The Steady State Approach*, Chapman & Hall, London, 1991.
8. Townsend, W. E. and Blankenship, L. C., Methods for detecting processing temperatures of previously cooked meat and poultry products—a review, *J. Food Prot.* 52, 128–135, 1989.
9. Townsend, W. E. and Blakenship, L. C., Enzyme profile of raw and heat processed beef, pork and turkey using the "APIZYM" system, *J. Food Sci.* 52, 511–512, 1987.
10. Stadler, J. W., Smith, G. L., Keaton, J. T., and Smith, S. B., Lactate dehydrogenase activity as an endpoint heating indicator in cooked beef, *J. Food Sci.* 62(2), 316–320, 1997.
11. Whitaker, J. R., Enzymes: monitors of food stability and quality, *Trends Food Sci.* 2(4), 94–97, 1991.
12. Joslyn, M. A., Enzyme activity in frozen vegetable tissue, *Adv. Enzymol.* 9, 613–652, 1949.
13. Muftugil, N., The peroxidase enzyme activity of some vegetables and its resistance to heat, *J. Sci. Agricul.* 36, 877–880, 1985.
14. Williams, D. C., Lim, M. H., Chen, A. O., Pangborn, R. M., and Whitaker, J. R., Blanching of vegetables for freezing—which enzyme to choose, *Food Technol.* 40(6), 130–140, 1986.
15. Parker, M. L., Ng, A., Smith, A. C., and Waldron, K. W., Esterified phenolics of the cell walls of chufa (*Cyperus esculentus* L.) tubers and their role in texture, *J. Agricul. Food Chem.* 48, 6284–6291, 2000.
16. Rose, J. K. C., Cosgrove, D., Albersheim, P., Darvill, A. G., and Bennet, A. B., Detection of expansin proteins and activity during tomato fruit ontogeny, *Plant Physiol.* 123(4): 1583–1592, 2000.
17. Vamos-Vigyazo, L., Polyphenol oxidase and peroxidase in fruits and vegetables, *Crit. Rev. Food Sci. Nutr.* 15(1), 49–127, 1981.
18. Cary, J. W., Lax, A. R., and Flurkey, W. H., Cloning and characterization of cDNAs coding for *Vicia faba* polyphenol oxidase, *Plant Mol. Biol.* 20(2), 245–253, 1992.
19. Hind, G., Marsha, D. R., and Coughlan, S. J., Spinach thylakoid polyphenol oxidase: cloning, characterization, and relation to a putative protein kinase, *Biochem.* 34(25), 8157–8164, 1995.
20. Strothkamp, K. and Mason, H., Pseudoperoxidase activity of mushroom tyrosinase, *Biochem. Biophys. Res. Commun.* 61, 827, 1974.
21. Zawistowski, J., Biliaderis, C. G., and Eskin, N. A. M., Polyphenol oxidase, in *Oxidative Enzymes in Foods*, Robinson, D. S. and Eskin, N. A. M., Eds, Elsevier Applied Science, New York, 217–273, 1991.
22. Dogan, M., Arslan, O., and Dogan, S., Substrate specificity, heat inactivation and inhibition of polyphenol oxidase from different aubergine cultivars, *Intern. J. Food Sci. Technol.* 37(4), 415–424, 2002.

23. Schwimmer, S., Fifty one years of food related enzyme research, *J. Biochem.* 19(1), 1–25, 1995.
24. Whitaker, J. R., Lipoxygenase, in *Oxidative Enzymes in Foods*, Robinson, D. S. and Eskin, N. A. M., Eds, Elsevier Applied Science, New York, 217–273, 1991.
25. O'Halloran, G. R., Troy, D. J., Buckley, D. J. and Reville, W. J., The role of endogenous proteases in the tenderization of fast glycolysing muscle, *Meat Sci.* 47, 187–210, 1997.
26. Ashie, I. N. A., Simpson, B. K. and Smith, J. P., Mechanisms for controlling enzymatic reactions in foods, *Crit. Rev. Food Sci. Nutr.* 36(1/2), 1–30, 1996.

# 12

# Postharvest Chemistry

## 12.1 INTRODUCTION

The challenge for the fresh food sector is to maintain quality at affordable prices. There are two modes of quality loss in fresh foods (Figure 12.1). Chemical deterioration occurs independently of enzymes (Chapters 9 and 10). Echoing the classical one-gene one-phenotype concept, biological deterioration can be described by a **one-enzyme one-defect model** (Chapter 11). Deterioration is due to multiple defect reactions each of which is catalyzed by a single enzyme. For instance, fruit softening is due to the enzyme "pectinase." Discoloration is ascribed to polyphenol oxidase (PPO). Off-flavor is linked to products formed by lipoxygenase. This atomistic view is oversimplistic. The one-enzyme one-defect model arose from laboratory studies using single enzymes and substrates. Food deterioration due to aging and senescence is an extremely complicated phenomenon, which is only beginning to be unraveled using genetic tools.

A major quality loss affecting noncitrus fruit is browning catalyzed by PPO (Chapter 11). The chemistry of this enzymic browning is relatively well understood. However, the one-enzyme explanation for browning is insufficient to account for the behavior of intact fruit. PPO is thought to be located in the cell cytoplasm. To be precise, the enzyme is located on the thylakoid membrane of chloroplasts. In contrast, the substrate phenols for PPO are found in the cell vacuole. During physical injury, the cell vacuole and chloroplast become damaged thereby bringing PPO and its substrates into contact. Cutting, peeling or rough physical handling causes physical damage and contributes to browning. Some experimental observations are also consistent with PPO being produced as an inactive precursor (see below). The PPO zymogen requires limited digestion by a protease to switch it on.

Classical models for food deterioration do not address the issue of control or timing. Deterioration is a time dependent (kinetic) process. Another important aspect is **coordination**. Fruits such as bananas develop brown spots and soften catastrophically after ripening. How is this discoloration and loss of texture *coordinated*? Assuming that such processes are enzyme catalyzed, what is the signal for the observed changes? To answer these questions scientists are developing more sophisticated models for biological deterioration. The potential for quality loss exists at all levels of the food system. An estimated 50% of fresh foods can be rejected by retailers as substandard. The rate of deterioration is affected by agricultural practices including growing and harvesting methods, postharvest handling, processing, distribution, and retail.

In this final chapter we consider emerging models for fresh food deterioration. Quality loss is described in terms **postharvest chemistry** catalyzed by multiple enzymes and regulated by external and internal signals.[1,2] Reactions taking place within fresh foods are organized into metabolic pathways. Deteriorative changes arise partly as a result of the loss of metabolic **regulation** and **homeostasis**. These changes are relatively rapid for animal derived foods.

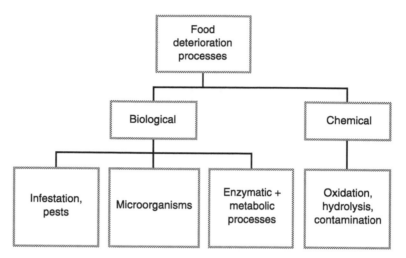

**Figure 12.1**  Food deterioration occurs via biological and chemical processes.

**Table 12.1**
**Food quality indices affected by biological deterioration**

| Quality attribute | Example |
|---|---|
| Sensory | Color, flavor (taste, smell, odor), texture (feel, consistency) |
| Nutritional | Biological value, mineral, and vitamin levels |
| Functional | Long-term health effects, probiotic components |
| Safety | Microbiological, levels of toxins, chemical contaminants |
| Storage time | Time for optimum storage |

Plant-based foods deteriorate more slowly probably in response to changes in **gene expression**. Biological deterioration also occurs due to **senescence** which is programmed decline following growth and maturation. Increasing knowledge of biological deterioration is being applied in the field of **genetic engineering**. Indices of food quality are listed in table 12.1.

## 12.1.1  HARVESTING

A whole plant is harvested by removing it from the soil. Harvesting fruits and vegetables means removing these organs from the donor plant, which is a source of nutrients, water, and hormones. Fish are harvested by removal from water, leading to asphyxiation. Livestock slaughtering practices include stunning, severing both carotid arteries, and decapitation. According to the Federation of European Veterinarians, slaughtering animals without prior stunning is unacceptable as this (1) increases the time to loss of consciousness, (2) causes additional pain due to exposed wound surfaces, aspiration of blood and rumen contents, and (3) requires additional restraints.[3] The carcasses are usually hung to drain free of blood, which otherwise acts as a medium for bacterial growth.

Harvesting is quickly followed by death of the whole organism. Most higher level functions are easy to disrupt by trauma. However, the line between life and death is not easy to define for excised organs, tissues or cells, which remain alive (after harvest) if refrigerated or bathed in a fluid with an appropriate pH and ionic strength. Plant cuttings and seeds remain viable for extended periods and can be regrown into new plants. Animal organs can be preserved for the purpose of transplantation.

Harvesting also disrupts homeostasis, which is a process that maintains a constant internal environment in the healthy organism. Without the blood circulation there is no transfer of oxygen and nutrients to the tissues. Cell reactions continue to utilize local reserves of starch or glycogen. Metabolic end-products from cells accumulate. Decapitation severs the spinal cord and prevents voluntary muscle action though spasmodic involuntary contractions continue. Physiological processes that normally operate under autonomic control (e.g., peristalsis) are also affected. The reticuloendothelial system responsible for immune function becomes impaired allowing invasion by extraneous bacteria. Harvesting fruits, leaves or whole plants interrupts the translocation of water and the reverse transfer of nutrients. Evaporation of water from the stomata continues, as does the breakdown of starch reserves.

### 12.1.2 POSTHARVEST METABOLISM

Metabolism is the totality of enzyme reactions occurring within a living cell. According to the **metabolic model**, food deterioration is a consequence of primary metabolism, which refers to cellular reactions directed at energy generation. **Catabolic** reactions result in the breakdown of fuel compounds in the cell into simpler products with the release of heat or chemical energy as ATP (adenosine triphosphate). Secondary metabolism consists of **anabolic** processes for the synthesis of low and high molecular weight compounds from simple building blocks. There is considerable scope for improving food quality by intervening in these reactions.

Metabolic reactions are organized into specific **biochemical pathways**. Each pathway is catalyzed by a group of enzymes. The two important primary metabolic pathways are **glycolysis** and the **Krebs cycle**. In a low oxygen environment, glycolysis transforms glucose to ethanol within plants cells. In animal tissue glycolysis transforms glucose to lactic acid. In the presence of oxygen, the products of glycolysis are carbon dioxide, water, and a much greater yield of energy. Secondary metabolism leads to the synthesis of cell micro-constituents (vitamins, colors, flavors) or macromolecules (proteins, polysaccharides, and lipids). Many incidental products of metabolism, including alcohols, aldehydes, and acids have a profound effect on food quality (Table 12.2).

---

**Table 12.2**
**Classification of metabolism**

| Description | Comments |
|---|---|
| *Primary metabolism* | Reactions for energy generation |
| Glycogenolysis | Breakdown of glycogen stored in muscle |
| Glycolysis | Main reaction for glucose breakdown, also known as fermentation, reactions lead to alcohol (plants) or lactic acid (animals) |
| Tri-carboxylic acid cycle | Oxygen requiring cyclic reactions that degrade sugars to carbon oxide and water, yielding 36 moles of ATP per mole of glucose metabolized. Reactions follow glycolysis provided oxygen is available |
| *Secondary metabolism* | Reactions for biosynthesis |
| Gluconeogenesis, photosynthesis, protein synthesis | Main pathways for synthesis of glucose, starch, and proteins |
| Shikimic acid pathway | Products include, benzenoid ring compounds, tryptophan, tyrosine, phenylalanine, flavanoids, anthocyanins |
| Acetate molanate pathway | Pathway utilizes acetate molecules for synthesis of fatty acids |
| Mevalonic acid pathway | Synthesis of isoprenoid monomers, carotenoids, pro-vitamin A, lycopene, cholesterol |

---

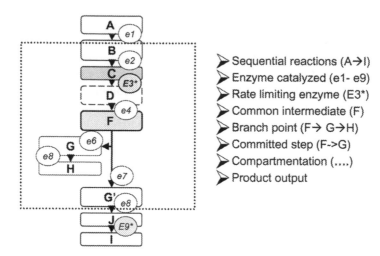

**Figure 12.2** A schematic diagram showing the organization of metabolic pathway for converting compound A to H and I via sequential reactions catalyzed by specific enzymes. *Regulatory enzymes that catalyze rate-limiting steps.

## 12.2 METABOLIC DESIGN AND ENGINEERING

The most notable metabolic pathways in living cells are listed in Table 12.2. Key design features of metabolic pathways are shown in Figure 12.2. This diagram will be used to discuss how different pathways are controlled. Reactions A → B, B → C are sequential and each step is catalyzed by a unique enzyme ($e_1$–$e_9$). Without a specific enzyme the corresponding reaction step will not occur. Cells can employ specific inhibitors to block particular enzymes in a metabolic pathway. The result is a build-up of compounds upstream from the blocked step.

In Figure 12.2, a branch point occurs at a **common intermediate** (F). This component can be channeled along the reaction sequence F → G → H or F → G′ → J depending on conditions in the cell. The reaction appearing immediately after a branch point is called a **committed step** because it leads to a well-defined product. One or more **rate-limiting** reactions control the rate at which material flows through a pathway. Regulatory enzymes catalyze the rate-limiting reaction and are usually subject to sophisticated feed-back (or feed-forward) control processes. The regulatory enzymes are sensitive to the concentration of one or more signaling compounds in a pathway. Consequently, the rate of material flux through a pathway can be turned up or down in response to changes in the concentration of **signaling compounds** — usually products from the metabolic pathway. Finally, metabolic pathways are **compartmentalized**. The sequence of reactions forming a pathway usually occur within well-defined cell compartments. To a rough approximation, different cell compartments correspond to different organelles (cf. lysosome, mitochondria, endoplasmic recticulum). The largest compartment of all is the cytoplasm bounded by a lipid cell membrane.

Food technologists and biotechnologists are seeking ways to control specific metabolic pathways. For instance it may be advantageous to enhance the formation of particular flavor compounds, vitamins or colors. However, before this can be done successfully, there is a need to identify the rate-limiting enzyme responsible for a pathway. Processes leading to fruit ripening are the result of *several* metabolic processes. A key rate-limiting step appears to involve two enzymes associated with the production of ethylene. **Genetic engineering** methods allowed the deletion of genes for ethylene production. Plants with reduced ethylene production exhibit delayed ripening. Fruits produced by such plants do not ripen unless exposed to external supplies of ethylene. As discussed later, this and other forms of **metabolic engineering** require painstaking work to elucidate factors affecting **flux** through the relevant pathway.

## 12.2.1 FLUX THROUGH METABOLIC PATHWAYS

Chemical reactions proceed to equilibrium within **closed systems** such as simple vessels (beakers, test-tubes or batch reactors) or enclosed spaces. At equilibrium the forward and reverse reaction rates are equal (Eq. 12.1). The equilibrium constant is defined as usual: $K_{eq} = [B]_{eq}/[A]_{eq}$.

$$\|A \rightleftharpoons B \rightleftharpoons C \rightleftharpoons D\|. \tag{12.1}$$

Living cells are **open systems**. Equilibrium is not attained in an open system owing to the continual "influx" of material and their removal out of the system (Eq. 12.2). We refer to the **mass action ratio** $(K_{mar})$ for the conversion $A \rightarrow B$, which is the ratio of observed concentrations $(K_{mar} = [B]_{ob}/[A]_{ob})$. In general, $K_{mar}$ may or may not be the same as $K_{eq}$ for the *same reactions measured in a closed system*. The ratio of concentrations at the source and sink also defines $K_{mar}$. Despite the absence of equilibrium conditions, the concentrations of reactants will appear constant over time, i.e., steady-state conditions will be reached provided the rate of influx and efflux of material are equal.

$$\text{Source} \rightarrow \|A \rightarrow B \rightarrow C \rightarrow D\| \rightarrow \text{Sink}. \tag{12.2}$$

Flux leads to a deviation from equilibrium. Likewise, a system in equilibrium is a closed system and cannot exhibit flux. For an open system, the extent of the deviation from equilibrium is measured by the **disequilibrium constant** $(\rho)$.

$$\rho = \frac{\text{mass action ratio}}{K_{eq}}, \tag{12.3}$$

where $\rho \neq 1.0$ is the condition for flux. The net disequilibrium constant for the reactions $A \rightarrow B$, $B \rightarrow C$, $C \rightarrow D$ is found by multiplying values for $\rho$ for $A \rightarrow B$, $B \rightarrow C$, and $C \rightarrow D$ (Eq. 12.4).

$$\rho_{net} = \prod_{i=1}^{n} \rho_i. \tag{12.4}$$

How can the preceding ideas be applied to metabolic pathways? Consider the eleven reactions of glycolysis (Table 12.3). With slight modification, this metabolic pathway is involved in alcholic fermentation by yeast. A slightly different version of the glycolysis pathway also converts glucose to lactic acid. This occurs as muscle cells metabolize glycogen as fuel, in the presence of limited amounts of oxygen, during the conversion of muscle tissue to meat. For red blood cells, glycolysis is the sole pathway for generating energy. The steady-state concentrations of intermediates are shown in column 2 of Table 12.3. In columns 3–5 are values for the mass action ratio, Gibbs free energy change, and (closed system) $K_{eq}$ values for reactions, $1 \rightarrow 2$, $2 \rightarrow 3$, $3 \rightarrow 4$, etc. The last column of Table 12.3 shows the disequilibrium constant values calculated using Eq. 12.3. Recall that $K_{eq}$ and $\Delta G$ ($= -RT \ln K_{eq}$) values are for reactions within a closed system.

It is noteworthy that $\rho \ll 1.0$ controls step(s) in a metabolic pathway[a]. From the last column of Table 12.3, $\rho \approx 0$ for three reactions: (1) glucose $\rightarrow$ glucose-6-phosphate, (2)

**Table 12.3**

**The 11 reactions of glycolysis transform glucose to lactic acid in red blood cells***

| Compound | Concentration (μM) | MAR | $\Delta G$ (cal/mol) | $K_{eq}$ | $\rho$ |
|---|---|---|---|---|---|
| 1. Glucose | 5000 | – | – | | |
| 2. Glucose-6-phosphate | 83 | 0.01660 | −4000 | 8.80E+02 | **0.00002** |
| 3. Fructose-6-phosphate | 14 | 0.16867 | 4000 | 1.14E−03 | 148.40 |
| 4. Fructose-1,6-diphosphate | 31 | 2.21429 | −3400 | 3.18E+02 | **0.0070** |
| 5. Dihydroxyacetone phosphate | 138 | 4.45161 | 5730 | 6.06E−05 | 73506.63 |
| 6. Glyceraldehyde-3-phosphate | 18.5 | 0.13406 | 1830 | 4.50E−02 | 2.98 |
| 7. 3-Phosphoglycerate | 118 | 6.37838 | 1500 | 7.87E−02 | 81.07 |
| 8. 2-Phosphoglycerate | 29.5 | 0.25000 | 1006 | 1.82E−01 | 1.38 |
| 9. Phosphoenolpyruvate | 23 | 0.77966 | 440 | 4.74E−01 | 1.64 |
| 10. Pyruvate | 51 | 2.21739 | −7500 | 3.32E+05 | **0.00001** |
| 11. Lactate | 2900 | 56.86275 | −6000 | 2.61E+04 | 0.0022 |

$\rho$ (net) = **0.000011**

* Based on Table 16.1 and $\Delta G$ values from Lehninger, A. L., *Biochemistry* (2nd edition), 423–433, 1975.

fructose-6-phosphate → fructose-1,6-diphosphate, and (3) phosphoenolpyruvate → pyruvate. Let us move from red blood cells to muscle cells, which also depend on glycolysis for much of their energy during periods of low oxygen supply. For muscle cells the rate-controlling step is the conversion of glycogen to glucose-6-phosphate. The importance of this reaction in the control of meat quality is described in Section 12.5.5. Finally, the disequilibrium constant for any single reaction can be unfavorably small so long as the net value for $\rho$ favors flux.

## 12.2.2 CONTROLLABILITY AND SENSITIVITY

Metabolic flux varies with enzyme concentration, pH, substrate concentration, and temperature. A graph showing the rate of flux with changing temperature has a gradient equal to the **controllability** (K) of a pathway as a function of temperature. In a similar way we can determine controllability by other variables. Also of interest is the sensitivity of enzymatic steps and whole pathways. **Sensitivity** is the fractional change in flux produced by a change in enzyme concentration.

## 12.2.3 REGULATION OF FLUX

The rate-limiting enzyme determines the rate of flux and pathway characteristics such as sensitivity and controllability. Within intact cells **regulatory enzymes** are controlled via a range of mechanisms (Table 12.4).

Allosteric enzymes possess another site other than the active site. Regulatory molecules activate or inhibit rate-limiting enzymes by binding to allosteric sites. In this way, cells achieve **negative feedback** control or **positive feed-forward** control of the rate-limiting enzyme. Allosteric modulators are cell metabolites produced downstream or upstream from the regulatory enzyme. Positive or negative allosteric modulators can fine-tune the rate of flux through the rate-limiting step in a pathway. Enzyme control by **covalent modification** involves

[a] We can calculate disequilibrium constant my measuring the observed concentrations in a cell and comparing this with the equilibrium constant for a bench-top reaction that is allowed to reach equilibrium. Values of $K_{eq}$ can also be readily calculated from standard data from chemistry handbooks.

**Table 12.4**
**Regulation of material flux through metabolic pathways**

| Process | Comment |
|---|---|
| Allosteric control | Enzyme is remotely controlled, switched on or off by a cell metabolite (e.g., ATP, AMP) |
| Covalent modification | Enzyme is activated or inhibited via covalent attachment of a phosphate group catalyzed by another enzyme called a protein kinase |
| Zymogen activation | An inactive pro-enzyme is activated by limited hydrolysis by a protease |
| Transcription control | The concentration of enzyme is increased by genetic control; expression of genes and increased protein synthesis |
| Compartmentation | Reaction is controlled by a physical (usually) membrane barrier between the enzyme and its substrate and the rate of influx to and from a specific compartment |

From: Stryer, L., *Biochemistry*, 4th edition, W. H. Freeman & Co., New York, 765–767, 1995, and Chapter 30, "Recurring motives in metabolic regulation."

the attachment of a phosphate group to an enzyme hydroxyl group, usually serine or threonine. Phosphorylation introduces an extra charge in the form of a phosphate group that causes a change in enzyme structure and activity.

Metabolic control via **zymogen activation** is possible where a rate-limiting enzyme is produced as an inactive precursor. Polyphenol oxidase, chymosin, and many digestive enzymes are synthesized as inactive zymogens and are later activated by limited proteolysis. **Transcriptional control** operates at the genetic level. Genes can be switched on at different times during the cell cycle though the exact mechanism is uncertain. Gene expression leads to high levels in messenger RNA, which is transcribed to enzyme protein. The concentration of any enzyme depends on their rate of synthesis and breakdown. **Compartmentation**, which is the separation of substrate and enzyme within different cellular compartments, is another method regulating metabolic reactions. A well-known example of compartmentation is the effect of bruising on enzymatic browning. The enzyme PPO discolors fruit as a result of mishandling, bruising or general physical abuse. Poor handling damages cell membranes, which separate PPO from its natural substrate. These are natural phenols normally sequestered within the plant cell vacuole. Oxidation of phenols catalyzed by polyphenol oxidase leads to brown pigments.

In the preceding discussion, we touched on the rapidly advancing area of **metabolic control analysis (MCA)** or **metabolic flux analysis** (MFA) which are attempts to describe the behavior of multi-enzyme systems. These topics have found direct application by industrial micro-biologists and those interested in enhancing the productivity of metabolic pathways in microorganisms. Recently, commercial computer software has been produced that enables the application of MFA for the production of useful food agents by fermentation or cell tissue culture. So far, applications of MFA to more complex food materials appears to be lacking[b].

## 12.3   SENESCENCE AND FRESH FOOD QUALITY

Senescence refers to a host of irreversible changes in living cells, normally ending with death. These are part of the normal developmental processes including growth, maturation,

---

[b] For a general introduction to MCA the reader is referred to the following website: http://dbk.ch.umist.ac.uk/mca_home.htm and associated links.

**Table 12.5**
**Natural and induced senescence**

| Natural senescence | Stress induced senescence |
|---|---|
| Leaf abscission in autumn (fall) | Acidity, alkalinity |
| Fruit ripening | Extreme cold or heat |
| Normal aging | Harvest stress (immature organs) |
| | Metal ions, toxins |
| | Nutrient deficiency |
| | Pathogen attack |
| | Physical wounding |
| | Salt stress |
| | Water stress (flood or desiccation) |

reproduction, decline, and death. For instance, aging is considered natural, genetically programmed, and not wholly avoidable. Recently, a distinction was made between **natural and induced senescence** (Table 12.5). Natural senescence is the result of normal aging whereas induced senescence is the result of exposure to stress (Table 12.5). According to recent modifications to the senescence model, some processing operations can produce changes in gene expression that ultimately affect food quality.[4]

## 12.4  MEMBRANE SENESCENCE

A lipid bilayer membrane surrounds living cells and is vital for their normal function. Drastic changes in cell membrane structure and function will have adverse effects on the quality of fresh foods. Membrane deterioration during senescence and its effects on food quality are considered here. The structure of animal and plant cell membranes consist of a **phospholipid bilayer**. Membrane proteins are embedded, island-like, within a phospholipid continuous phase. Some very large proteins span the width of the cell membrane, passing from the outside to the inside of the cell compartment. Membrane **fluidity** allows **lateral diffusion** and collision of proteins and enzyme subunits necessary for proper cell function. A low membrane **microviscosity** is necessary for fluidity determined by lipid composition: high unsaturated fatty acid content increases membrane fluidity. High levels of cholesterol, a normal component of cell membranes, increases membrane rigidity.

Normally, the plasma membrane serves as a "smart" barrier separating the cell interior from the external medium. Both plant and animal cells are **semi-permeable** and will only allow the diffusion of water or small, neutral, molecules into the cell. Glucose, and other neutral molecules, enters the cell after binding with membrane proteins in a process called **facilitated transport**. The lipid cell membrane is not permeable to charged species. Therefore, ionic molecules enter the cell with the aid of membrane-located **active transport** proteins. Unlike simple diffusion, or facilitated transport, active transport requires metabolic energy in the form of ATP. Active transport enables the uptake of ions, against the prevailing concentration gradient, from a region of low concentration to a region of high concentration. Receptor proteins, which bind to signaling molecules such as hormones, are also located on the cell membrane.

**Cell organelles** are also surrounded by a phospholipid bilayer. The **endoplasmic recticulum**, sometimes studded with **ribosomes**, acts as an intracellular passageway for the transfer of proteins headed for export via the **Golgi vesicles**. Within plant cells, the large central **vacuole** is surrounded by a single bilayer membrane, the **tonoplast**. The vacuole contains an aqueous phase with dissolved acids, phenolic compounds, and secondary metabolites. The vacuole contents are sequestered and kept separate from the rest of plant cell.

**Table 12.6**
**A summary of cell membrane types and their function**

| Types of cell membranes | Function |
|---|---|
| Microsomal membrane fraction | |
|    Plasma membrane | Active barrier, selective transfer of solute |
|    Endoplasmic recticulum | Post-translational modification of proteins |
|    Golgi apparatus | Packaging, export of cellular constituents |
|    Tonoplast | Membrane around the plant cell vacuole |
| Mitochondria membrane | Electron transport, and ATP generation |
| Chloroplast membrane | Electron transport, photolysis of water |

Membrane fractions isolated from cell fragments are listed in Table 12.6. The different membrane fractions possess different buoyancy. It is relatively simple to separate different classes of cell membranes using **density gradient centrifugation** or **liquid–liquid two phase partitioning**.[5] The **microsome fraction** consists of fragments of plasma membrane, endoplasmic recticulum, Golgi vesicles, and the tonoplast. The collective function of the microsome fraction is homeostasis, i.e., the maintenance of a constant intracellular environment. By contrast, the membranes associated with the **mitochondria** and the electron transport chain underpin the generation of ATP. There are also plant cell membrane fragments from the chloroplast and chromoplasts. Chloroplasts are concerned with the generation of intracellular stores of energy via photosynthesis. Clearly, the integrity of cell membranes will affect cell viability and senescence.

### 12.4.1  ROLE OF FREE RADICALS

Membrane deterioration is partly the result of attack by **free radicals**. Evidence for a free radical mechanism stems from studies using antioxidants and direct measurement of free radical levels in aging cells. Baker and co-workers reported that sodium benzoate and rhizobixine, two free radical scavengers, delayed fruit ripening. Antioxidants also increase the lifespan of cut carnations.[6] The steady state concentration of the superoxide radical, measured by reacting with tiron (1,2-dihyroxybenzyne-3,5-disulfonic acid), increased for aging tissue. In that study, tiron semi-quinone radical was monitored using electron spin resonance (ESR).

### 12.4.2  CELL MEMBRANE ORGANIZATION

During senescence the microsome fraction changes its **composition**. The ratio of cholesterol: phospholipid rises due to the breakdown of phospholipids by **phospholipase** $D$ (PLD).[7] The role of PLD in membrane senescence is described further in Section 12.4.5. There is also also an increase in the level of lipid **hydroperoxides** due to enzymatic and nonenzymatic oxidation (Section 12.4.6).

### 12.4.3  MEMBRANE FLUIDITY AND MICROVISCOSITY

Senescing membranes show decreased **fluidity** and increased **microviscosity**. The changes were observed by measuring the rotational motion of paramagnetic membrane probes using ESR.[8] Membrane microviscosity was determined using fluorescence polarization measurements. The decrease in fluidity coincided with increased levels of superoxide radicals.[9] This suggested that decreases in membrane microviscosity was due to peroxidation of membrane lipids, perhaps catalyzed by lipoxygenase. Carbonyl products from lipid peroxidation could cross-link membrane lipids thereby increasing membrane viscosity.

### 12.4.4 PHASE CHANGES AND NON-BILAYER AGGREGATES

The fraction of membrane lipids existing in a gel-like state increases with senescence. The formation of a **crystalline** lipid phase was measured using wide-angle X-ray diffraction and from melting point determination.[10,11] The crystallization of biological membrane lipids may result in **rigidification** of the lipid bilayer.

Another structural change for senescing membrane lipids is their transformation into non-bilayer, vesicle or **reverse micelle**, structures. Formation of such structures would explain increases in membrane **permeability** that accompany senescence. Aging membranes become increasingly permeable to ionic species (hydrogen, sodium, potassium or calcium ions). Reverse micelles increase ionic transfer. Peroxidation products formed from fatty acids and phospholipids could also act as calcium ionophores accounting for increases in membrane permeability that occur with aging.[12,13]

### 12.4.5 PHOSPHOLIPASE AND MEMBRANE SENESCENCE

There is increased hydrolysis of phospholipids during membrane senescence. The reaction is catalyzed by phosholipases.[14] Studies using **Arabidopsis** as a model show that plant cells possess multiple forms of phospholipase D (PLD). Each isoenzyme is encoded by a separate gene. At least five PLD isoforms ($\alpha$, $\beta$, $\gamma_1$, $\gamma_2$, $\gamma_3$) have been identified, which differ in their requirement for $Ca^{2+}$ ions, pH activity characteristics, response to lipid substrates, and transcriptional control in response to stress (Table 12.7).

The most abundant phospholipase in *Arabidopsis* appears to be PLD$\alpha$, which requires 20–100 mM of $Ca^{2+}$ for activity. The $Ca^{2+}$ requirement reaches physiological ($\mu$M) levels when PLD$\alpha$ is assayed under acidic (pH 4.5–5.0) conditions likely to form under stress. The substrate requirement of PLD$\alpha$ includes phosphatidylcholine (PC), phosphatidylethanolamine (PE) and phosphatidylglycerol (PG). PLD$\beta$ and PLD$\gamma$ had more restrictive substrate requirements, with activity restricted to substrate mixtures having more than 50% PE.

PLD is inhibited by lysophosphatidylethanolamine (LPE), a hydrolysis product of PE, leading to slowed senescence. Cabbage and castor bean tissue PLD were both inhibited by LPE via an uncompetitive process. In contrast, lysophosphatidylserine, lysophosphatidylcholine or

---

**Table 12.7**

**Effect of stress on the expression of phospholipase D mRNA in *Arabidopsis* leaves**

| Stress | PLD$\alpha$ | PLD$\beta$ | PLD$\gamma$ |
|---|---|---|---|
| Control | ++ | − | − |
| NaCl (1%) | +++ | + | + |
| Mannitol (0.6 M) | +++ | ++ | ++ |
| AlCl$_3$ | ++ | ++ | - |
| CdCl$_2$ | ++ | ++ | ++ |
| Peroxide | +++ | ++ | − |
| Abscisic acid | ++ | + | ++ |
| 4 °C | ++++ | − | − |

Adapted from Wang et al., *Biochem. Soc. Trans.* 28(6), 813–816, 2000.

lysophosphatidylglycerol had no inhibitory effect. Leaves treated with LPE had a higher chlorophyll content, lower rate of respiration, and lower ethylene production. Tomato treated with LPE showed lower ethylene production and lower degrees of electrolyte leakage compared with nontreated samples.[15]

Zang and co-workers used **antisense technology** to suppress PLDα activity and showed that this enzyme is important for senescence in *Arabidopsis*. Genetically modified leaves with low PLDα activity show delayed yellowing, decreased electrolyte leakage, greater photosynthetic activity, and higher levels of chlorophyll and phospholipids compared to the wild-type plant. For wild-type leaves, treatment with abscisic acid and ethylene stimulated PLDα expression.[16] The PLDα antisense genes did not affect normal plant development. Other PLD isoforms also undergo changes in the rate of transcription in response to stress.

### 12.4.6 LIPOXYGENASE AND MEMBRANE SENESCENCE

Membrane aging is associated with lipid oxidation catalyzed by lipoxygenase. Formation of lipid hydroperoxides facilitates membrane accessibility by phospholipase. A fall in the antioxidant systems of the cell during aging could also play a role in membrane deterioration. Two antioxidant enzymes, superoxide dismutase (SOD) and glutathione peroxidase, which help to protect cells from free radicals, may contribute indirectly to membrane senescence (Figure 12.3). The level of SOD rises in cells exposed to high oxidative stress.

To summarize, membrane deterioration is an integral part of the **senescence syndrome**. Programmed aging leads to membrane destabilization resulting from (1) an altered phospholipid:cholesterol ratio, (2) increased formation of gel-like semi-crystalline membrane phases, (3) decreased membrane fluidity and increased microviscosity, and (4) formation of

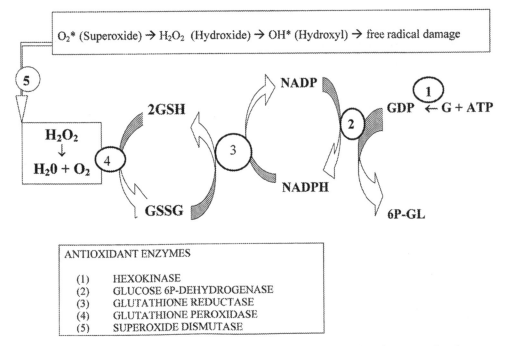

**Figure 12.3** Enzymic antioxidant systems for cells. The tripeptide glutathione cycles between the reduced (GSH) and oxidized (GSSG) states and so helps control the level of potentially damaging oxygen radical species.

non-bilayer vesicle structures. Such alterations in membrane structure and function affect compartmentation—the ability of membranes to divide the cell into distinct functional regions. The changes in ionic environment within cells lead to activation of previously masked enzymes and an increase in the degradation of proteins and lipids. Membrane senescence involves phospholipase and lipoxygenase, which together catalyze the peroxidation of unsaturated lipids. The present summary is merely a sketch of membrane senescence. Cell aging and programmed cell death is an area of active research.

## 12.5 PATTERNS OF FRUIT GROWTH AND RIPENING

Fruits exhibit sigmoidal growth (apples) or a double sigmoidal growth (pears). In the former case the dry weight follows an S-shaped increase with time. These patterns of growth are determined by the underlying cell cycle: lag phase, growth, cell division, and expansion. Growth is followed by maturation then senescence. From a biology standpoint, **ripening** increases the attractiveness of fruits to bats, birds, and other vectors for **seed dispersal**. From a food technology viewpoint premature ripening is undesirable. It would be useful to (1) delay ripening during storage and transportation and (2) control the nature of ripening to produce higher quality foods.

### 12.5.1 CLIMACTERIC AND NONCLIMACTERIC FRUIT

Climacteric fruit was discovered in the 1920s by Kid and West. These are fruits which show increases in oxygen consumption and increased production of ethylene at the onset of ripening.[17] Externally applied ethylene shortens the time till the onset of ripening. The increase in oxygen uptake is permanent or irreversible. With ethylene addition, the *maximum* rate of oxygen uptake by climacteric fruit is unchanged compared to control fruit with no added ethylene.

A second group of fruits are **nonclimacteric** fruits,[18] which ripen without an increase in oxygen uptake and without an obvious peak in ethylene production. Externally added ethylene produces an immediate rise in oxygen uptake but the effect is transient and stops if ethylene is removed (Figure 12.4). For both climacteric and nonclimacteric fruits, ripening is hastened by added ethylene. In contrast, ripening is delayed by continually removing naturally produced ethylene from the surroundings of fruit by flushing with air. Examples of climacteric and nonclimacteric fruit are listed in Table 12.8.

The biochemical difference between climacteric and nonclimacteric fruit proved illusive until recently. The discovery of **ripening inhibited (*rin*) mutants** of climacteric fruit pointed to a genetic basis for climacteric behavior. Unlike normal climacteric tomato, *rin* tomato shows nonclimacteric traits. There is no respiratory increase or increased ethylene production at the onset of ripening.

### 12.5.2 FRUIT STRUCTURE AND COMPOSITION DURING RIPENING

Fruit ripening produces changes in food **ultrastructure**, color, texture, and flavor. Electron microscope observations show that cells from ripened fruit have reduced numbers of **chloroplasts**; those that remain contain lower levels of chlorophyll. Thalakoid membranes inside chloroplasts become disrupted. In this manner, chloroplasts are transformed into aged versions, called **chromoplasts**. Earlier literature indicated that plant cell organelles (mitochondria, Golgi structure, and ribosomes) were unaffected by ripening. More recent information points to a number of subtle biochemical changes in cell membranes during senescence

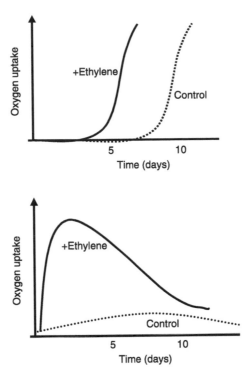

**Figure 12.4**　Changes in the pattern of oxygen uptake for climecteric fruit (top) and nonclimacteric fruit (bottom) with and without (control) the addition of ethylene.

## Table 12.8
### Examples of climacteric and nonclimacteric fruit

| Climacteric fruit | Nonclimacteric fruit |
| --- | --- |
| Apple, apricot, avocado | Blueberry |
| Banana, breadfruit | Grape |
| Fig | Honeydew melon |
| Cantaloupe melon | Java plum |
| Pear | Lemon, lime, grapefruit, orange |
| Mango | Olive |
| Tomato | Pineapple |
| | Strawberry |

(Section 12.4). A summary of changes of fruit and vegetables during senescene is given in Table 12.9.

There is a color change from green to yellow during fruit ripening. This is due to chlorophyll breakdown and unmasking of pre-existing **lycopene** molecules within chromoplasts. Synthesis of lycopene occurs via the **mevalonic acid pathway** that forms 5C isoprene units (Table 12.1). The red, pink, purple, and blue coloration of some fruit is due to **anthocyanins** found within plant cell vacuoles. The dyes vary in color according to the ambient pH and minor variations in chemical structure.

Gross changes in the texture of plant-based foods is due to the breakdown of plant cell walls by glucosidases, e.g., exo- and endo-polygalacturonase (pectinase) and $\beta(1 \rightarrow 4)$D-glucanase (cellulase). Pectin acts as an intercellular cement. Therefore, pectin hydrolysis reduces cell–cell adhesion and leads to tissue softening. The enzyme pectinase hydrolyses

**Table 12.9**

**A summary of changes in fruit and vegetables during senescence**

| Change | Explanation |
| --- | --- |
| Sensory change | |
|   Color | Decrease in green color, increase in yellow, red |
|   Texture | Softening, drying, hardening |
|   Flavor | Increase in volatiles, increase in sweetness, reduced sourness |
| Physiological and cellular change | Loss of chlorophyll and chloroplast structure |
| | Degradation of plant cell wall |
| | Altered membrane structure, reduced fluidity, increased permeability, release of vacuole contents |
| Biochemical changes | |
| | Altered metabolism, starch hydrolysis |
| | Loss of RNA |
| | Increase proteolysis, net loss of protein |

isolated pectin molecules in a test-tube. Nevertheless, hydrolysis of cell wall pectin does not seem to be the rate-limiting step for softening within intact fruit. It is probable that compartmentation limits pectinase access to cell wall located pectin.

Flavor compounds associated with ripened fruits include organic acids, sugars, and a host of volatile compounds (Table 12.10). There is a fall in pH during fruit ripening. Depending on the fruit, considerable amount of acids (e.g., malic acid, citric acid) are sequestered within the vacuole. Starch is hydrolyzed to glucose, which is then transformed into sucrose. The volatiles from fruits are low molecular weight esters formed from 5-carbon alcohols (e.g., isoamyl alcohol, isopentyl alcohol) and butanoic acid. A great many flavors are aldehydes and ketones produced as secondary breakdown products from hydroperoxides, which are, in turn, products of lipid oxidation.

### 12.5.3 Correlations Between Ripening Processes

That oxygen uptake increases for climacteric fruits due to the higher energy requirement for ripening is not supported by recent evidence. A coincidence between the ripening phenomena does not prove that these events are causally related. The following observations indicate that respiration is not strictly related to ripening:

(1) Nonclimacteric fruit ripen without an increase in respiration.
(2) Short exposure to ethylene induces ripening without a respiration increase.
(3) Protein synthesis inhibitors (e.g., cycloheximide) prevent ripening without affecting the climacteric event.
(4) The onset of endogenous ethylene production does not always coincide with oxygen uptake. Ethylene production precedes oxygen uptake by 10 days or a few hours for honeydew melon and banana, respectively.

### 12.5.4 Ethylene as a Plant Hormone

Ethylene is the principal internal agent controlling ripening. A build-up of ethylene stimulates further ethylene production leading to a positive feedback loop. Ethylene is directly involved in the ripening process as evidenced by the following:

(1) Addition of ethylene induces ripening in climacteric and nonclimacteric fruit.
(2) Hypobaric (low atmospheric pressure) storage delays fruit ripening and reduces the accumulation of ethylene within plant tissue.

**Table 12.10**
**Fruit flavor compounds**

| Products | Fruit flavor compounds |
| --- | --- |
| Apple | *n*-hexanal, ethyl butyrate, 1-propyl propionate, 1-butyl acetate, *trans*-2-hexenal, ethyl 2-methylbutyrate, 2-methylbutyl acetate, 1-hexanol, hexen-1-ol, *trans*-2-hexen-1-ol, hexyl acetate, esters, alcohols, aldehydes, ketone, acid, including hexanal, ethyl 2-methyl butyrate |
| Apple juice | Esters, alcohols, aldehydes, ketones, acids, terpenes, sulfur-containing compounds |
| Banana | Alcohols, esters, including amyl acetate, isoamyl acetate, butyl butyrate, amyl butyrate |
| Cherry | Benzaldehyde |
| Cranberry | Benzaldehyde; benzyl and benzoate esters |
| Grape | Methylanthranilate |
| Grapefruit | 1-*para*-menthane, 8-thiol nootkatone, limonene, psoralen nootkatone |
| Mango | Cyclohexane, methylcyclohexane, hydrocarbon, dimethylcyclohexane, ethylcyclohexane, 1,1-diethoxyethane, ethanol, alpha-pinene, 1-methylpropan-1-ol, toluene, alpha-fenchene, camphene, hexanal, an ethyl butenoate, butan-1-ol, beta-pinene, sabinene, xylene, car-3-ene, myrcene, alpha-phellandrene, 3-methylbutan-1-ol, limonene, beta-phellandrene, gamma-terpinene, *para*-cymene, alpha-terpinolene, *cis*-hex-3-en-1-ol, 2-furfural, ethyl octanoate, alpha-copanene, beta-caryophyllene, ethyl decanoate, sabinyl acetate, alpha-humulene, ethyl dodecanoate |
| Orange | Acetaldehyde, ethanol, limonene, ethyl esters, linalool; alpha-terpincol |
| Orange peel oil | *n*-hexanal, *n*-heptanal, 6-methyl-5-hepten-2-one, *n*-nonanal, *trans*-limonene oxide, *cis*-limonene oxide, octyl acetate, citronellal, *n*-decanal, *n*-undecanal, neral, geranial |
| Peach | Ethyl acetate, dimethyl disulfide, *cis*-3-hexenyl acetate, methyl octanoate, ethyl octanoate, 6-pentyl alpha pyrone, gamma decalactone |
| Peach | Benzaldehyde, benzyl alcohol, gamma-caprolactone, gamma-decalactone |
| Pear | Esters of 2,4-decadienoic acid, especially the esters of ethyl, *n*-propyl, *n*-butyl |
| Pineapple | 4-methoxy-2,5-dimethyl-2(*H*)-furan-3-one, 2-propenyl hexanoate, sesquiterpene hyrocarbons, 1-(*E,Z*)-3,5-undecatriene, 1-(*E,Z,Z*)3,5,8-undecatetraene. 2-propenyl *n*-hexanoate ethyl, *para*-allyl phenol; gamma-butyrolactone; gamma-octalactone; acetoxyacetone; methyl esters of beta-hydroxybutyric, beta-hydroxyhexanoic acids |
| Strawberry | Ethylcinnamates, methylcinnamates, 2,5-dimethyl-4-hydroxy-3(2*H*)-furanone, furaneol, furaneol-beta-glucoside, dimethyl-4-methoxy-3(2*H*)-furanone (mesifurane), methyl and ethyl acetates, propionates, butyrates |

Adapted from http://food.oregonstate.edu/faq/plant/fruit/fruit_flavor.html

(3) Inhibition of ethylene synthesis using a fungal toxin, rhizobitoxine (2-amino-4′(2′-amino-3′-hydroxypropoxy)*trans*-3-butanoic acid) reduces the symptoms of ripening (respiratory rise, texture loss) in apples and pears.

(4) Ethylene induces ripening in a low oxygen atmosphere where respiratory rise cannot occur.

(5) Different threshold concentrations of ethylene are needed to induce a range of ripening changes (Table 12.11).

The *effects* of ethylene on plants have been known indirectly since ancient Egyptian times. The ancient Chinese burnt incense in closed rooms to enhance the ripening of fruits. Leaks of gas from street lights affected the growth of plants. In 1901, Dimitri Neljubow showed that the

Table 12.11

Threshold concentrations of ethylene induce different ripening changes

| Fruit | Ripening process | Ethylene concentration ($\mu$l/L or ppm) |
|---|---|---|
| Melon | Texture loss | 0.1 |
| | Carotenoid formation | 0.1 |
| | Respiration increase | 3.0 |
| Pear | Texture loss | 0.08 |
| | Respiration increase | 0.46 |
| Tomato | Pectin loss | 5 |
| | Respiration increase | 0.1–0.2 |

active component was ethylene. In 1917, ethylene was shown to stimulate abscission and in 1934 it was reported that plants produced ethylene. In 1935, it was proposed that ethylene was the plant hormone responsible for fruit ripening. Ethylene is now known to have many other functions as well.

In 1965 a mechanism of ethylene action as a plant hormone was proposed. A hormone is any compound secreted at one site of an organism but which has its effect at a distant site. In this instance, the hormone happens to be gaseous. The effect of differing ethylene concentrations on ripening responses was described in terms of specific binding of ethylene to an unidentified receptor:

(1) The intensity of ripening effects increased with the concentration of ethylene added. At high concentrations of ethylene the ripening effects reached a maximum, indicating saturation.
(2) The concentration of ethylene required for 50% maximal response was approximately 0.6 nM.
(3) Results could be explained by specific binding of ethylene molecules with an unidentified receptor with affinity for compounds $R.C=CH_2$. Of a series of homologous alkenyl compounds, ethylene was the most strongly bound.
(4) Oxygen was required for binding ethylene to plant cell receptors. Recently, the ethylene receptor was cloned in *Arabidopsis*. The ethylene receptor protein had a hydrophobic N-terminal hormone binding site. Mutations in this region abolished ethylene binding and reduced the plant's sensitivity to this hormone. The hydrophilic carboxy terminal of the ethylene receptor appeared to be a histidine protein kinase. Thus, plant responses to ethylene appeared to involve a 2-component hormone receptor, not unlike the hormone receptor systems for hormones like adrenaline in animals (see later).

Ethylene is one of the five classes of plant hormones that include **abscisic acid** (ABA), **auxin** (indole acetic acid (IAA), **cotokynins** (zeatin), and **gibberellins** (GA). The functions of these hormones are summarized in Table 12.12. Abscisic acid occurs in green fruit, leaves, and stems where it facilitates closure of stomata during water stress. Auxins, cytokinins, and gibberellins promote cell division and elongation, and can be considered suppressors of senescence. Ethylene appears to be the plant hormone most clearly associated with promotion of ripening and senescence.

## 12.5.5 PRACTICAL CONTROL OF RIPENING BY EXTERNAL MEANS

Internal control of ripening by ethylene was described in the previous section. Exposing fruits to ethylene also provides a means for external control of ripening. Other technologies for external control of ripening include the use of (1) oxygen, (2) carbon dioxide,

**Table 12.12**

**Functions of plant hormones**

**Abscisic acid**

- Stimulates the closure of stomata (water stress increases ABA synthesis)
- Inhibits shoot growth and may promote growth of roots
- Induces seeds to synthesize storage proteins
- Inhibits the effect of gibberellins on stimulating *de novo* synthesis of $\alpha$-amylase
- Induces gene transcription for proteinase inhibitors during wounding, plays a role in pathogen defense

**Auxin**

- Stimulates cell elongation
- Stimulates differentiation of phloem and xylem
- Stimulates root initiation on stem cuttings and lateral root development in tissue culture
- Mediates the tropistic response of bending in response to gravity and light
- The auxin supply from the apical bud suppresses growth of lateral buds
- Delays leaf senescence
- Delays fruit ripening
- Promotes (via ethylene production) femaleness in dioecious flowers
- Stimulates the production of ethylene at high concentrations

**Cytokinin**

- Stimulates cell division
- Stimulates morphogenesis (shoot initiation/bud formation) in tissue culture
- Stimulates the growth of lateral buds-release of apical dominance
- Stimulates leaf expansion resulting from cell enlargement
- May enhance stomatal opening in some species
- Promotes the conversion of etioplasts into chloroplasts via stimulation of chlorophyll synthesis

*continued*

**Table 12.12**
**Continued**

**Ethylene**

$H_2C=CH_2$

- Stimulates the release of dormancy
- Stimulates shoot and root growth and differentiation (triple response)
- Stimulates leaf and fruit abscission
- Induction of femaleness in dioecious flowers
- Stimulates flower opening
- Stimulates flower and leaf senescence
- Stimulates fruit ripening

**Gibberellins**

- Stimulate stem elongation by stimulating cell division and elongation
- Stimulates bolting/flowering in response to long days
- Breaks seed dormancy in some plants which require stratification or light to induce germination
- Stimulates enzyme production ($\alpha$-amylase) in germinating cereal grains for mobilization of seed reserves
- Induces maleness in dioecious flowers (sex expression)
- Can cause parthenocarpic (seedless) fruit development
- Can delay senescence in leaves and citrus fruits

Adapted from the BBSRC (UK). http://www.plant-hormones.info/index.htm accessed Jan 2004.

(3) temperature, and (4) pressure. Increasing oxygen levels within the range 1–5% accelerates ripening. Therefore, reduced oxygen atmospheres (comprising mainly nitrogen) can be used for long-term storage of fresh fruits in warehouses. Elevated carbon dioxide levels (3–10%) inhibit ripening though the mode of action remains uncertain. Temperature is a common means for external control of ripening which occurs at 10–35°C. Care is needed with tropical and sub-tropical crops, which are susceptible to **chilling injury** during refrigeration. Finally, avoiding ethylene build-up in the atmosphere surrounding fruits is another way to control ripening. We must not forget also that addition of ethylene facilitates ripening. Such technologies are being applied to fruits for out-of-season consumption.

## 12.6 MOLECULAR BIOLOGY OF FRUIT RIPENING

Prior to the 1980s preservation methods for fresh foods (Section 12.5.5) were developed by empirical observation. Let us now consider the molecular biology era and some emerging technologies for ripening control.

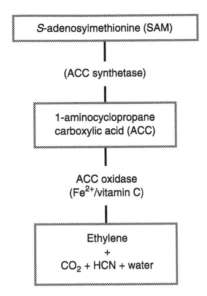

**Figure 12.5** Ethylene synthesis by ACC synthetase and ACC oxidase.

## 12.6.1 ETHYLENE SYNTHESIS

Young and Hoffman elucidated the pathway for ethylene synthesis in 1984. The metabolic pathway has two enzymes (Figure 12.5). The precursor for ethylene synthesis is *S*-adenosylmethionine (SAM) which is the storage form of methionine within cells. SAM is formed from methionine and ATP. The committed step for ethylene synthesis is the conversion of SAM to 1-aminocyclopropane 1-carboxylic acid (ACC) catalyzed by ACC synthetase (ACCS). Next, ACC is converted to ethylene by ACC oxidase, (ACCO). This enzyme contains iron ($Fe^{2+}$) in its active site and requires vitamin C for proper function.

## 12.6.2 REGULATION OF ETHYLENE SYNTHESIS

Ethylene biosynthesis is under **transcriptional control** (Table 12.3). The rate of ethylene formation is controlled by the quantities of two enzymes, ACCS and ACCO, present within cells. In the tomato, nine genes code for ACCS and three code for ACCO. The multiple genes are subject to **differential expression**. Some genes are switched in a time-dependent manner by developmental factors (aging). Other ethylene related genes are activated by external signals for senescence, including high temperatures, high salinity, wounding, and pathogen attack (Figure 12.5).[19] At the onset of ripening, ACCS and ACCO genes are read by the cell machinery to form messenger RNA. mRNA for the ethylene-producing enzymes is translated to form ACCS and ACCO. These enzymes are part of a group of about 19–25 ripening related proteins whose concentration increases at the point of ripening. It is noteworthy that ethylene is one of the signals for increasing ethylene synthesis (Figure 12.6).

## 12.6.3 MOLECULAR BASIS FOR THE CLIMACTERIC EFFECT

Climacteric and nonclimacteric fruit have fascinated plant biologists for many years. However, extensive studies failed to find consistent differences between these groups at the biochemical level. Observed differences between climacteric and nonclimacteric fruit were slight compared

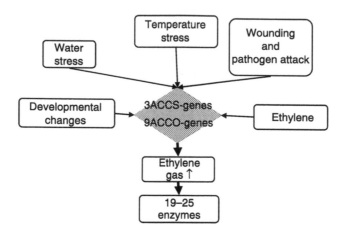

**Figure 12.6** Regulation of ethylene biosynthesis during senescence by natural (developmental) and external (stress) factors.

to the general differences between different species. Recent findings suggest that climacteric and nonclimacteric fruiting plants differ according to their ethylene synthesis machinery.

Two major systems exist for ethylene synthesis in plants. **System I** occurs in nonclimacteric and pre-climacteric fruit. Enzymes for system I are **constitutive**, meaning they are continually produced within fruits regardless of the ripening stage. The system II ethylene synthesis pathway is present only in climacteric fruit, and is **inducible**. System II ethylene synthesis can be switched on by a range of internal and external signals for senescence, and is sensitive to ethylene from system I.

During growth and maturation, pre-climacteric fruit as well as nonclimacteric fruit produce ethylene at a low rate controlled by system I. The genes for system I enzymes (ACCS and ACCO) are transcribed into mRNA and then translated into protein **constitutively**. Eventually, the concentration of ethylene reaches a threshold level necessary to induce ACCO and ACCS genes of system II. As a result, a burst of ethylene production follows which now switches on the host of ripening related enzymes including those associated with the climacteric effect.

### 12.6.4 ANTISENSE GENE TECHNOLOGY

From 1988 reports begun to appear describing the isolation of total mRNA from ripening tomato. mRMA was generally copied to form complementary DNA (cDNA) which is a more stable form of nucleic acid for storage. Comparing the cDNA from ripened and green tomatoes led to the identification of 19–25 fragments large enough to code for "genes." The ripening-related genes were designated TOM1–TOM25. Working from the base sequence and the corresponding amino acid sequence, TOM6 was identified as coding for the pectin-hydrolyzing enzyme, polygalacturonase.

TOM6 was subsequently isolated, and the placed in a reverse orientation within a vector and injected inside tomato cells growing in **tissue culture**. The cells were induced to form **callus** and then whole plants. These were allowed to self-pollinate to form fruits. Tomatoes from plants genetically modified using the so-called **antisense genes** showed a reduction in polygalacturonase mRNA. The fruits also had lower levels of enzyme activity compared to the wilde-type fruit. Interestingly, the texture of normal tomato and the GM tomato, carrying the polygalacturonase antisense gene, were the same. This finding was unexpected based on the assumption that fruit texture loss is due to pectin hydrolysis by pectinase. Curiously, GM tomato had some interesting characteristics: (1) reduced shriveling and splitting associated with over-ripening, (2) small but significant changes in firmness at later stages of ripening, and (3) increased paste viscosity.

GM tomato carrying an antisense gene for polygalacturonase is marketed as FLAVRSAVR by Calgene Ltd (USA). A second pectinase (pectin methyl esterase) has also been switched off using antisense gene technology. Here too, the GM tomato showed no large differences in texture compared to normal tomatoes. Apparently, the rate limiting step for the loss of fruit texture within *intact* fruit is not pectin hydrolysis. Alternatively, pectin is not hydrolyzed by pectinase within intact fruits. Enzyme access to the substrate may be restricted in intact tomato.[20] Actually, plant texture is a function of a hierarchy of structural features, including cell wall, composition, degree of cross-linking, lignification, and wall–wall adhesion within tissues.[21]

## 12.7 BIOCHEMICAL CHANGES IN MUSCLE FOODS

Muscle foods quality is determined by a number of variables that affect tissue pH; (1) pre-slaughter handling, (2) tissue glycogen reserves at slaughter, (3) carcass storage temperature, and (4) electrical stimulation. The conversion of muscle into meat is determined by the rate of metabolism of glycogen to form lactic acid during the postslaughter period. The conversion of muscle tissue into meat is discussed in this section.

### 12.7.1 MUSCLE STRUCTURE AND CONTRACTION CYCLE

Muscle foods include tissues from domesticated livestock (e.g., cow, sheep, goat, buffalo), fowl, and poultry, pigs, fish. Meat is essentially skeletal muscle. Offal derived from smooth muscle of the guts and intestines is widely consumed, as are products like sausages, black pudding, and haggis.

The muscle cell ranges from 1 cm to 1 m in length, has multiple nuclei, and is surrounded by a lipid bilayer membrane, the sarcolema. The usual complement of organelles (endoplasmic recticulum, Golgi system, lysozomes) is found within muscle cells; mitochondria (site of ATP synthesis) are especially prominent within highly aerobic tissue. A high concentration of **myoglobin**[c] is responsible for the coloration of red meat. In contrast, less intensely colored meat has low concentrations of myoglobin. A transparent membrane the **perimysium**, covers a bundle of muscle cell fibers. This same covering gives rise to **tendons**, which attach muscle to bone. Muscle tissue also contains blood vessels, and varying amounts of **cartilaginous** material.

The unit for muscle contraction is called the **sacomere**. Electron micrographs show that each sarcomere contains an array of thick **myosin** filaments. Each thick filament is surrounded by a hexagonal array of **actin**. During muscle contraction, the sarcomeres shorten as thick myosin filaments slide past the thin actin filaments. Tension arises within muscle due to repeated binding interactions between actin and myosin. In resting muscle, actin and myosin binding is prevented by two regulatory proteins, troponin and tropomyosin. Muscle

---

[c]Myoglobin is a muscle protein, which functions in oxygen binding and storage. It is structurally similar to hemoglobin from red blood cells, which functions in oxygen transport.

stimulation by nerve impulses involves the following steps:

(1) Depolarization of the muscle cell membrane. The small electrical potential normally present across the muscle cell membrane is disrupted causing the release of calcium ions into the cell interior.
(2) $Ca^{2+}$ ions bind to troponin causing this protein to change its shape. The conformation change allows actin–myosin binding and muscle contraction.

The energy for dissociating the actin and myosin comes from the hydrolysis of ATP. This occurs at the myosin head region which functions as an ATP-hydrolyzing enzyme (ATPase). Therefore, the energy demand for muscle contraction is used for relaxation. Some ATP is also used to power the active transport of $Ca^{2+}$ ions back into their intracellular stores after contraction. A rise in intracellular $Ca^{2+}$ levels serves as the signal for muscle contraction.

### 12.7.2  CALCIUM IONS AND ENERGY CONTRACTION COUPLING

Elevated intracellular $Ca^{2+}$ also triggers the breakdown of glycogen to provide the ATP for muscle contraction. Biochemists refer to the role of $Ca^{2+}$ in muscle **contraction–energy coupling**. See Section 12.5.6 for details of how $Ca^{2+}$ stimulates glycogen breakdown and ATP production.

### 12.7.3  CONVERSION OF MUSCLE TO MEAT

The conversion of muscle to meat occurs in three stages (Table 12.13). In the immediate postslaughter period blood circulation is interrupted. There is no transfer of oxygen from the lungs to the tissues. Immediately after death, intracellular levels of ATP remain high ensuring muscle relaxation and a soft texture. As ATP becomes depleted muscle tissue enters a state of **contraction–rigor**. Longer still, muscle texture loss occurs due to **tissue breakdown** by lysosomal proteases or **cathepsins** that become activated at low pH. The changes that accompany the conversion of muscle to meat are closely tied to the extent of postharvest metabolism within muscle. Muscle also contains two other protease systems that may contribute to texture loss: calcium activated **calpains** and **ATP/ubiquitin** dependent proteosomes.[22]

To ensure optimum meat quality, muscle pH should fall from the physiological value of about pH 7.3 to pH 5.3 over a 24-h period. Two deviations from optimal conditioning lead to sub-standard meat. A slow decrease of tissue pH to pH 6–6.5 produces low quality, close textured *DFD* **meat** (dry, dark, firm meat). The high pH of **DFD meat** makes this product more susceptible to microbial spoilage. DFD meat defect arises from low levels of muscle glycogen at the point of slaughter. Poor pre-slaughter practice (fasting, rough handling, transportation of live animals over long distances) increases the likelihood of DFD meat. Excessive stress or struggling during the pre-slaughter period reduces muscle glycogen levels.

A rapid decline in tissue pH ($\sim$0.2–0.22 pH units/h)[d] occurs for meat held near body temperature (35°C). The result is another defect called **PSE meat** (pale, soft, exudates meat). PSE meat has poor color, low water binding ability, and high drip loss. The analogous texture

---

**Table 12.13**

**Stages in the conversion of muscle to meat**

| Stage | Comments |
|---|---|
| Pre-rigor stage | Muscle ATP high, meat is soft |
| Rigor stage | ATP is depleted, stiff texture |
| Post-rigor stage | Softening or tenderizing phase due to action of proteases |

defect is found in fish. PSE meat is formed by the denaturation of muscle proteins by a combination of low pH and high storage temperature.

Storage at 1–5°C produces an undesirable increase in carcass rigidity termed **cold shortening**. Low temperature causes the release of $Ca^{2+}$ ions from intracellular stores. In the absence of ATP for moving $Ca^{2+}$ back into cellular stores, there is sustained muscle contraction leading to shortening and increased texture. Cold shortening can be avoided by holding meat at 15°C. Alternatively, scientists from New Zealand found that electrical stimulation of meat carcasses prevents cold shortening. The scientific explanation for this technology is uncertain. External electricity may function like electrical nerve impulses. Nerve impulses usually stimulate contraction and also stimulate the production of ATP by glycolysis. Apparently, cold shortening may be due to the failure of energy-contraction coupling (Section 12.7.2) in meat stored at low temperatures. Electrically stimulated meat showed a pH fall from pH 7.3 to pH 6.0 in about 3 h. By comparison, the pH falls by 1.3 units after 15.4 h for unstimulated meat.[23–25]

### 12.7.4 GLYCOGEN METABOLISM AND MEAT TECHNOLOGY

Glycogen is the sole fuel for muscle contraction. There is no supply of glucose to muscle cells in the absence of an intact blood circulation. Glycogen breakdown generates ATP and lactic acid as the end products. Post-mortem tissue pH and ATP levels are determined by quantities of glycogen stored within muscle and their breakdown rate. The net amount of glycogen stored in muscles (1% w/w) exceeds the total stored in the liver (~10% w/w) owing to the greater total mass of muscle.

The structure of glycogen is similar to amylopectin (cf. linear $\alpha(1 \rightarrow 4)$D-glucose backbone with $\alpha(1 \rightarrow 6)$ branch points). However, glycogen is more highly branched and significantly more soluble. The open structure of glycogen makes this material easily hydrolyzable by enzymes. Glycogen is converted to glucose-6-phosphate (G-6P) by a process called phosphorolysis. This is similar to hydrolysis except that phosphate (rather than water) is the major reactant (Eq. 12.5). The phosphorolysis of glycogen is catalyzed by the enzyme **glycogen phosphorylase**. This is undoubtedly the single most important enzyme affecting meat quality.

$$G_{(n/\text{residues})} + Pi \rightarrow \text{glycogen}_{(n-1\ \text{residues})} + \text{G-6P}. \tag{12.5}$$

G-6P is then metabolized to lactic acid via glycolysis (Table 12.2).

The rate-limiting step for glycogen breakdown in muscle tissue is the reaction catalyzed by **glycogen phosphorylase** (Eq. 12.5). This large enzyme has two subunits each of which weighs about 97,000 Da. Each Subunit has a substrate-binding site for glycogen and another (allosteric) binding site for AMP, ATP or G-6P. Also noteworthy is the presence of a unique hydroxylamino acid (serine 14) side chain ($CH_2$–OH). Glycogen phosphorylase is under two forms of control: (1) metabolic control via an **allosteric mechanism**, and (2) electrical or hormonal control acting via **covalent modification** (Table 12.4). AMP produced from ATP hydrolysis in postmortem muscle, binds to glycogen phosphorylase and activates this enzyme (Eq. 12.5) to stimulate glycogen degradation. In an intact animal, falling ATP levels can be readily replenished. Lactic acid produced in the muscle is transported to the liver where it is converted to glucose and then sent back to the muscle. High levels of ATP and G-6P have a sparing effect on glycogen. Both ATP and G-6P exert **negative feedback** control on glycogen

---

[d]At temperatures of 10–35°C there is a 2-fold increase in the rate of fall of tissue pH (from 0.05–0.22 pH per hour) for every 10°C change in storage temperature.

phosphorylase. In contrast, glycogen synthetase is stimulated by high levels of ATP and G-6P leading to synthesis of glycogen. **Electrical stimulation** activates glycogen phosphorylase.[26] Recall that electrical signals from the nerves produce an influx of $Ca^{2+}$ ions into muscle cells. High $Ca^{2+}$ levels stimulate muscle contraction. At the same time, $Ca^{2+}$ also activates the enzyme PK-A (protein kinase A). PK-A switches on glycogen breakdown by activating the glycogen degrading enzyme, glycogen phosphorylase. The covalent activation of glycogen phosphorylase by protein kinase is summarized in Eq. 12.6. Electrical stimulation, whether it occurs via nerve impulses or externally applied electrical voltages, enhances glycogen breakdown.

$$\text{inactive.Ser14–OH} + \text{ATP} \xrightarrow{\text{PK-A}} \text{active.Ser14–O. PO}_4 + \text{ADP.} \tag{12.6}$$

### 12.7.5   Lactic Acid Formation and the Quality of Muscle Foods

The breakdown of glycogen and glycolysis leads to the formation of **lactic acid**. The accumulation of this metabolite is responsible for the decline in the postmortem tissue pH. The build-up of acidity disrupts **lysosomes**, which are cell organelles that contain digestive enzymes. Lysosomal enzymes are able to hydrolyze cell polymers. Failure of containment releases lysosomal enzymes, which then perform wholesale digestion of cell contents. A fall of intracellular pH is the signal for quality change in muscle foods.

### 12.7.6   Kinase Cascades and Signal Transduction

The stress-hormone **adrenaline** is secreted by adrenal glands located at the kidneys. From there it is transported to the muscle tissues via the circulating blood. Adrenaline binds to a hormone receptor located on the plasma membrane releasing a G-protein associated on the intracellular side of the receptor. The G-protein dissociates from a hormone receptor and binds to adenylate cyclase which now produces cAMP. This so-called **second messenger** is a positive allosteric modulator for protein kinase. The final step in the activation of glycogen breakdown by adrenaline is shown in Eq. 12.6.

Interestingly, ethylene appears to function in a manner similar to adrenaline. Recall that ethylene is the plant hormone that controls senescence processes including ripening. Ethylene binding to plant cell membrane receptors leads to the generation of an unknown intracellular signal. A series of 5–7 protein kinases (PK1-5) then become activated. PK5 modifies a DNA transcription protein, which controls ripening related genes. Given that plants and animals evolved quite separately, it seems remarkable that signal transduction involves similar **protein kinase cascades**.

## 12.8   POSTHARVEST VERSUS POSTMORTEM CHANGES

Postharvest changes in plant foods are readily described in terms of **senescence**. In contrast, postmortem changes in animal foods occur rapidly due to **metabolic processes**. Plant and animal tissue appear to be fundamentally different in their response to stress. Recall from biology that animals respond to stress by flight or fight, i.e., via a metabolic response. Plants are incapable of movement unless by growth, involving **gene expression**. Likewise, plant and animal foods appear to respond differently to processing stress.

Glycolysis leads to the generation of ethyl alcohol, which causes off-flavor in stored fruits. The hydrolysis of starch in plant foods produces a loss of dry matter during storage.

Manipulation of $O_2$ and $N_2$ levels can be carefully managed in order to avoid excessive formation of ethanol. Oxygen is needed for ethylene synthesis. The Krebs cycle leads to the formation of large amounts of organic acids that accumulate in the plant cell vacuole in sour fruits and vegetables. During the postharvest period, there is no evidence for the build-up of lactic acid in plants.

Aging affects meat quality. **Maturation** leads to a general **toughening** of meat, which is the result of an increase in the size and amounts of muscle fiber. There is some loss of elasticity due to an increase in the number of cross-links in collagen. Clearly, the senescence model can also be applied to meat products. Nevertheless, most meat livestock do not live past an optimum age. Maturation is usually followed by a **decline in feed conversion**. There are also clear textural differences between veal and beef, or mutton dressed as lamb.

# References

1. Cornish-Bowden, A., Metabolic control analysis in biotechnology and medicine, *Nature Biotechnol.* 17, 641–643, 1999.
2. Cornish-Bowden, A. and Luz Cárdenas, M., From genome to cellular phenotype—a role for metabolic flux analysis? *Nature Biotechnol.* 18, 267–268, 2000.
3. Federation of Veterinarians of Europe, Slaughter of animals without prior stunning: a position paper, 2002. Avilable on http://www.fve.org/papers/pdf/aw/position /papers/02_104.pdf (accessed January 2004).
4. King, G. A. and O'Donoghue, E. M., Unraveling senescence: New opportunities for delaying the inevitable in harvested fruit and vegetables, *Trends Food Sci. & Technol.* 6(12), 385–389, 1995.
5. Yoshida, X. X., Uemur, M., Niki, T., Sakai, A., and Gusta, L. V., Partition of membrane particles in aqueous two polymer phase systems and its practical use for purification of plasma membranes from plants, *Plant Physiol.* 72, 105–114, 1983.
6. Baker, J. E., Wang, C. Y., Lieberman, M., and Hardenburg, R., Delay of senescence in carnation by a rhizobixine analogue and sodium benzoate, *Hort. Sci.* 12, 38–39, 1977.
7. Paliyath, G. and Droillard, M. J., The mechanism of membrane deterioration and disassembly during senescence, *Plant Physiol. Biochem.* 30, 789–812, 1992.
8. Borochov, A., Halevy, A. H., and Shinitzky, M., Increase in micro viscosity with ageing in protoplast plasmalema of rose petals, *Nature (London)*, 263, 158–159, 1976.
9. Mayak, S., Legge, R. L., and Thompson, J. E., Superoxide radical production by microsomes membranes from senescing carnation flowers: an effect on membrane fluidity, *Photochemistry* 22, 1375–1380, 1983.
10. McKersie, B. D. and Thompson, J. E., Lipid crystallization in senescent membranes from cotyledons, *Plant Physiol.* 59, 803, 1977.
11. McKersie, B. D. and Thompson, J. E., Phase properties of senescing plant membranes, role of neutral lipids, *Biochim. Biophys. Acta* 550, 48–58, 1979.
12. Shaw, J. M. and Thompson, J. E., Effect of phospholipid oxidation products on trans-bilayer movement of phospholipid in single lamella vesicles, *Biochemistry* 21, 920–927, 1981.
13. Serhan, C., et al., Phosphotidate and oxidized fatty acids are calcium ionophores. Studies using arsenzo II in liposomes, *J. Biol. Chem.* 256, 2736–2741, 1980.
14. Wang, X., Wang, C., Sang, Y., Zheng, L., and Qin, C., Lipids and signaling: phospholipase-mediated pathways, *Biochem. Soc. Trans.* 28(5), 813–816, 2000.
15. Ryu, S. B., Karlsson, H., Ozgen, M., and Palta, J. P., Inhibition of phospholipase D by lyso-phosphatidylethanolamine, a lipid-derived senescence retardant, *Proc. Natl. Acad. Sci. USA* 94, 12717–12721, 1997.
16. Fan, L., Zheng, S., and Wang, X., Antisense suppression of phospholipase D alpha retards abscisic acid- and ethylene-promoted senescence of postharvest *Arabidopsis* leaves, *Plant Cell* 9(12), 2183–2196, 1997.
17. Laties, G. G., Kidd, F., West, C., and Blackman, F. F., The start of modern post harvest physiology, *Postharvest Biol. Technol.* 5, 1–10, 1994.

18. Biale, J. B., Growth, maturation and senescence in fruit, *Science*, 146, 880–888, 1964.
19. Morgan, P. W. and Drew, M. C., Ethylene and plant responses to stress, *Physiologia Plantarum* 100, 620–630, 1997.
20. Picton, S., Gray, J. E., and Grierson, D., The manipulation and modification of tomato fruit ripening by expression of antisense RNA in transgenic plants, *Euphytica* 85, 193–202, 1995.
21. Waldron, K. W., Smith, A. C., Parr, A. J., Ng, A., and Parker, M. L., New approaches to understanding and controlling cell separation in relation to fruit and vegetable texture, *Trends Food Sci.* 8, 213–220, 1997.
22. Dransfield, E. and Sosnick, A. A., Relationship between muscle growth and poultry quality, *Meat Science* 78, 743–746, 1999.
23. Bendall, J. R., Electrical stimulation of rabbit and lamb carcasses, *J. Sci. Food Agricul.* 27, 819, 1976.
24. Tarrant, P. J. V. and Sherrington, J., An investigation of ultimate pH in the muscles of commercial beef muscle, *Meat Sci.* 4, 287, 1980.
25. Hornikel, K. O., Roncalés, P., and Hamm, R., The influence of temperature on shortening and rigor onset in beef muscle, *Meat Sci.* 8, 221–224, 1983.
26. Chrystall, D. B. and Devine, R. F., Electrical stimulation, muscle tension, glycolysis in bovine sternomandibulus muscle, *Meat Sci.* 2, 49, 1978.

# Index